Membranes in Gas Separation
and Enrichment

Special Publication No. 62

Membranes in Gas Separation and Enrichment

The Proceedings of the Fourth BOC Priestley Conference, sponsored by BOC Limited and organised by the Royal Society of Chemistry in conjunction with the University of Leeds

Leeds, 16th—18th September 1986

ROYAL
SOCIETY OF
CHEMISTRY

CHEM
7185297

British Library Cataloguing in Publication Data

BOC Priestley Conference *(4th : 1986 : Leeds)*
Membranes in gas separation and enrichment :
the proceedings of the Fourth BOC Priestley
Conference, sponsored by BOC Limited and
organised by the Royal Society of Chemistry
in conjunction with the University of Leeds,
Leeds, 16th–18th September 1986.—
(Special publication; no. 62).
1. Gases—Separation   2. Membrane
separation
I. Title   II. Royal Society of Chemistry
III. University of Leeds   IV. Series
660.2'842   TP242
ISBN 0-85186-676-X

Published by The Royal Society of Chemistry,
Burlington House, London, W1V 0BN

Printed in Great Britain by
Whitstable Litho Ltd., Whitstable, Kent

TP 242
B 621
1986
CHEM

Introduction

The Fourth BOC Priestley Conference was held in the University
of Leeds on 16th-18th September, 1986, with the title
'Membranes in Gas Separation and Enrichment'. This extremely
timely conference considered not only the fundamental
physico-chemical aspects of the membrane separation of gases
but also its increasing application to industrial gas
separation processes and the concomitant economic aspects.
In addition one session was devoted to the history of the
development of gas separation processes.

Two major highlights of the conference were The Priestley
Lecture and the BOC Centenary Lecture. The BOC Centenary
Lecture was the opening lecture of the conference and was
presented by Professor P. Meares (University of Exeter) who
outlined, in an excellent lecture, the fundamental mechanisms
of transport of small molecules in solid polymers. At the
beginning of this lecture the BOC Centenary Medal was presented
to Professor Meares by Dr. J. W. Barrett.

The Priestley Lecture was given by Dr. E. Brian Smith
(University of Oxford) who spoke on 'The Science of Deep-Sea
Diving - Observations on the Respiration of Different Kinds
of Air'. This fascinating lecture, chaired by Lord Dainton,
was given to a large audience including a number of school
children amongst whom were the winners and runners-up of
the BOC Priestley Schools Competition.

The main scientific programme covered three general areas:
the physical chemical basis of the membrane separation of
gases, the synthesis of membrane materials and the production
of the membranes themselves, and the applications of membrane
separation to industrial processes. In the first lecture
D. Tillmann (Linde, FRG) outlined the conditions which needed
to be met in order for membrane systems to compete with
existing methods of gas separation, this being followed by
an outline of some technical and economic aspects of gas
separation by means of membranes by U. Werner (University
of Dortmund, FRG). It became clear that what is still required
is a 'magic' membrane with properties superior to all existing
membranes. Papers by R. Krishna (Indian Institute of
Petroleum) and by J. Barrie (Imperial College, London) covered
fundamental aspects of separation processes.

An important group of papers dealt with the choice of
membranes. Two papers, J. E. McIntyre (University of Leeds)
and C. A. Smolders (Twente University, Netherlands) dealt
with the production and properties of hollow fibres, whilst
two others, I. K. Ogden (British Petroleum) and S. A. Stern
(University of Syracuse and Gas Research Institute, USA)
dealt with new membrane materials.

The major industrial application of membrane separation
at the present time is the separation of hydrogen although
there are some other important separations and these were
discussed in four papers by I. W. Backhouse (Monsanto, Europe),
J. G. O'Hair (British Gas), I. Hakuta (National Chemical
Laboratory for Industry, Japan) and R. L. Schendel (Fluor
Technology, USA). The recovery of helium from diving gas
by membranes was presented by K. V. Peinemann (GKSS, FRG).
Three presentations by K. B. McReynolds (Dow Chemicals,
Europe), T. Nakagawa (Meiji University, Japan), and C.-L.
Lee (Dow Corning and Gas Research Institute, USA) dealt with
the difficult separation of oxygen and nitrogen by membranes,
but they showed that air separation systems are possible
and discussed its potential for the future.
The final two papers were by W. van der Eijk (Commision
of the European Communities) and D. L. MacLean (BOC, USA).
The first dealt with the EEC BRITE programme whilst the latter
dealt with the structure of gas separation technology in
a final global overview paper.
The historical section arranged by Professor Colin Russell
(Open University) dealt with 'Thomas Graham and Gaseous
Diffusion' (M. Stanley, Wednesbury), 'The Discovery of Gas
Liquid Chromatography' (A. T. James, formerly of Unilever,
Sharnbrook), Low Temperature Separation of Gases (G. G.
Haselden, University of Leeds), and Pressure Swing Adsorption
Methods (N. F. Kirkby, University of Surrey). These lectures
provided a fascinating insight into the development of the
subjects.
The international range of speakers from industry and
Universities resulted in a highly attended conference of
very considerable interest. I am most grateful to all the
lecturers for the time and effor they devoted to their
presentations, to the chairmen, and to those who exhibited
posters. I should like to acknowledge the contributions
of the members of the Programme Committee whose efforts
resulted in such an excellent scientific programme. I also
thank Dr. J. W. Barrett and members of the Organising Committee
for their support, and Dr. J. F. Gibson and his staff for
their efficient administrative arrangements.

Professor A. Williams, University of Leeds
Chairman, Programme Committee.

Contents

BOC Centenary Lecture
Fundamental Mechanisms of Transport of Small
Molecules in Solid Polymers   1
   By P. Meares

Conditions Which Need to be Fulfilled by Membrane
Systems in Order to Compete with Existing Methods
of Gas Separation   26
   By W. Baldus and D. Tillmann

Some Technical and Economic Aspects of Gas
Separation by Means of Membranes   43
   By U. Werner

A Unified Theory of Separation Processes Based
on Irreversible Thermodynamics   64
   By R. Krishna

The Transport of Water in Polymers   89
   By J. A. Barrie

Polymers for Gas Separation Membranes   114
   By I. K. Ogden, R. E. Richards, and A. A. Rizvi

Production of Porous Hollow Polysulphone
Fibres for Gas Separation   130
   By G. C. East, J. E. McIntyre, V. Rogers, and
   S. C. Senn

Hollow Fiber Gas Separation Membranes: Structure
and Properties   145
   By Z. Borneman, J. A. van't Hoff, C. A. Smolders,
   and H. M. van Veen

Structure-Permeability Relationships in Silicone
Polymer Membranes   158
   By S. A. Stern, V. M. Shah, and B. J. Hardy

Thomas Graham and Gaseous Diffusion   161
   By M. Stanley

The Discovery of Gas-Liquid Chromatography: A
Personal Recollection   175
   By A. T. James

The Story of Law Temperature Gas Separation   201
By G. C. Haselden

Pressure Swing Adsorption Methods   218
By N. F. Kirkby

Priestley Lecture
On the Science of Deep-sea Diving: Observations
on the Respiration of Different Kinds of Air   230
By E. B. Smith

Recovery and Purification of Industrial Gases
Using Prism^R Separators   265
By I. W. Backhouse

The Use of Membranes in the Japanese 'C$_1$
Chemistry Programme'   281
By T. Hakuta, K. Haraya, K. Obata, Y. Shindo,
N. Ito, and H. Yoshitome

Applications of Membrane Technology in the
Gas Industry   291
By B. W. Laverty and J. G. O'Hair

Separation of Acid Gas and Hydrocarbons   311
By R. L. Schendel

The Recovery of Helium from Diving Gas with
Membranes   329
By K.-V. Peinemann, K. Ohlrogge, and H.-D. Knauth

Generon Air Separation Systems - Membranes in
Gas Separation and Enrichment   342
By K. B. McReynolds

Polyacetylene Derivatives as Membranes for
the Separation of Air   351
By T. Nakagawa

Silicone Polymer Membranes for Air Separation   364
By C.-L. Lee, H. L. Chapman, M. E. Cifuentes,
K. M. Lee, L. D. Merrill, K. C. Ulman, and
K. Venkataraman

The Structure of Gas Separation Technology   382
By D. L. MacLean

Posters Presented at the Conference   399

Address Preceding the Toast to Joseph Priestley
at the Priestley Dinner on 18th September 1986   401
By R. V. Giordano

BOC Centenary Lecture
Fundamental Mechanisms of Transport of Small Molecules in Solid Polymers

By P. Meares

DEPARTMENT OF CHEMICAL ENGINEERING, UNIVERSITY OF EXETER, NORTH PARK
ROAD, EXETER EX4 4QF, UK

Background to Gas Separation by Polymer Membranes

While membranes have been used for many years to carry out
separations in the liquid phase by dialysis and electrodialysis,
their industrial use in the separation of mixtures of gases and
vapours is little more than ten years old. This may seem strange
because already in 1831 Mitchell[1] recorded the rates of escape
of ten gases from natural rubber balloons and showed there was
a range of 100-fold between carbon monoxide and ammonia. He
noted too that the rapidly permeating gases were absorbed by
rubber to a considerable extent.

Formal study of liquid phase permeation in synthetic
membranes began only some thirty years later with the work of
Thomas Graham. He also reported gas transport through natural
rubber[2] and discussed the solution-diffusion mechanism of
transport. Graham introduced the notion that, internally,
rubber was more like a liquid than a solid; this despite the
fact that the molecular nature of rubber was entirely unknown
at the time. He demonstrated also that when air permeates
natural rubber the permeate was sufficiently enriched in oxygen
to relight a glowing splint.

Despite this early recognition of the separating potential
of rubber membranes for air, it has taken 120 years to develop
viable practical devices for recovering oxygen and nitrogen by
diffusion through polymer membranes. In the intervening period
much research has been done on the penetration of many gases
and vapours singly through a variety of polymers. This work has
led to a considerable degree of understanding of the molecular
mechanisms by which small molecules are absorbed by and transported in polymers. These mechanisms form the main subject of

1

this lecture.

When the scope of the field was widened beyond rubber and
air, the possibilities of several useful gas separations became
apparent,[3] but early attempts to exploit them failed because of
the low permeabilities of the membranes made by currently
available techniques from polymers with adequate permselectivity.

The practical feasibility of such gas separations came
about quite suddenly through developments in membrane production
for another liquid phase separation process, namely reverse
osmosis. In the beginning the asymmetric membranes (in which an
ultrathin layer, 100 nm thick or less, of dense polymer is
supported on a porous and readily permeable layer about 1000 times
thicker than this) developed for reverse osmosis were stable
only when wet and lost their asymmetric structure on drying.
Once techniques had been discovered to produce membranes in the
dry state while retaining their asymmetric structure,[4] the way
was opened for rapid advance in the gas separation field also.

Although the development of such ultrathin membranes,
asymmetric or composite in structure, and their incorporation
in spiral wound and hollow fibre permeation modules has gone a
long way towards solving the problem of the low intrinsic
permeabilities of useful membrane polymers, their selectivities
are still too low to carry out a number of desirable separations
in a single pass. Staging or cascading is possible but adds to
capital cost through reduced production rate and to running cost
through increased energy consumption.

The search for polymer membranes more selective for
particular separations is now being pursued vigorously, especially
in U.S.A. and Japan and there is speculation about being able to
tailor polymers to meet specific separation requirements. As
usual, effective technology can rest only on sound scientific
understanding which in this case is of the interactions between
polymer chemistry, membrane structure and the permeability to
particular gases and vapours. Hence this choice of topic to
open the BOC Priestley Conference on Membranes in Gas Separation
Enrichment.

The Solution-Diffusion Mechanism

The mechanism of permeation of gases in rubber, introduced
by Graham and still regarded as essentially correct for other
polymers also, is that the membrane contains no permanent pores
through which the gas flows. Instead, the polymer acts as a
homogeneous solvent in which the gas dissolves to a well-defined
equilibrium concentration determined by the natures of the polymer
and the gas and by temperature and pressure.

When the gas at one face of the membrane is at a higher
pressure than at the other, it dissolves to a higher concentration
at that face of the membrane and diffuses, as a result of the
random molecular heat motion of the gas molecules and segments
of the polymer chains, down a concentration gradient to the low
pressure face where it evaporates so as to maintain the correct
equilibrium concentration at the low pressure side.

It should be noted that the dissolution of gas in a polymer
is a typical partition phenomenon with a thermodynamic equilibrium
coefficient and a heat and entropy of mixing involving contributions from both components.

The molecular interactions of typical simple gases with the
polymer are small and the dissolved concentrations at moderate
pressures are low. Then one expects Henry's law to hold with a
constant solubility coefficient S. With organic vapours larger
amounts are absorbed by the polymer and the isotherms are not
linear. The solution-diffusion model is still applicable but the
mathematical treatment has to be altered, usually by substituting
the Flory-Huggins equation [5] instead of Henry's law. This
adaptation is particularly effective at the high volume fractions
of polymer in membrane systems.

The diffusion of gas within the polymer is governed by the
thermo-molecular kinetics. The jumping frequency of the gas
molecules is far higher than that of the polymer segments which
are consequently rate controlling. These jumping frequencies
are strongly temperature dependent and they involve, like
viscous flow, a free volume of activation. The availability
of this free volume increases with the thermal expansion

of the material.

Because simple gases dissolve to only low concentrations
their presence in the polymer scarcely affects its molecular
motions. Diffusion is then governed by a constant diffusion
coefficient D and the most straightforward form of Fick's
law can be applied. Since the permeation flux J is a function
of the concentration of gas in the polymer and the rate at which
it is transported, both D and S enter into the flux
equation. When both are constants the relation is

$$J_s = ADS\, \Delta p/h \qquad\qquad (1)$$

where A is the membrane area, h its thickness, Δp the
difference in gas pressure between the two sides and J_s is
measured once a steady flow rate has been established.

To achieve even a minimum understanding of the physico-
chemical factors governing the permeability of a particular gas
and polymer it is necessary to determine D and S separately.

Where permeation in asymmetric membranes is considered,
the solution-diffusion mechanism applies only to permeation in
the dense or active layer of the membrane. Membranes prepared
by the various phase inversion methods [6] have complex graded
structures and it is not possible to characterize them with an
independently measured value of the thickness h . Scientific
studies of gas permeation are thus preferably carried out on
homogeneous films. Most work to date has been on isotropic
polymer films although orientation is known to bring about major
changes in permeability and so does a gradient of composition
through a dense film when the sorption isotherms are non-linear.

Experimental Methods

With only few exceptions, two basic methods of observation
have been used. They are permeation through the polymer sample
from one gas phase to another and measurements on the rate of
sorption or desorption of the gas or vapour from or into a
single gas phase. Within these two classes of measurement a

great variety of techniques has been adopted. These have
recently been covered in an excellent review.[7]

   The diffusion of gases has most frequently been studied
by following permeation through a flat film of the polymer from
one vessel to another. Provided measurements are made in both
the transient state that follows a change in the gas phases
outside the membrane and in the steady state ultimately reached,
the data can be analyzed to give the diffusion and sorption
coefficients separately. This may be understood by considering
a particularly simple experimental procedure illustrated in
Figure 1. In this, the membrane separates two closed vessels
which can be evacuated.

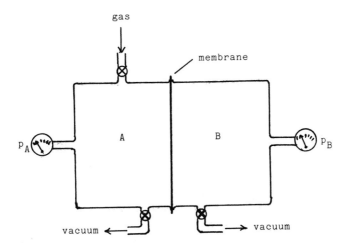

Figure 1   Gas permeation apparatus (schematic)

   The gas under study is introduced suddenly into A and its
pressure and temperature held constant thereafter. Gas which
permeates the membrane collects in B the volume of which is
known. It is held at the same temperature as A. Either the
pressure in B is measured as a function of time or its volume
is adjusted so as to maintain a constant low pressure of gas in

the vessel. In either case the pressure in B has to be kept far
less than that in A . We shall consider the former case.

    If at zero time a pressure p_A is established in A, gas
will dissolve in the membrane at the side facing A. It is
assumed, though without much clear evidence, that sorption
equilibrium is established instantly at that face. Dissolved
gas begins to move through the membrane and after an interval
emerges into B which is initially evacuated. The release of
gas into B occurs slowly at first and then increasingly rapidly
until a steady rate is eventually set up. The time course of
the pressure in B, p_B, is shown in Figure 2.

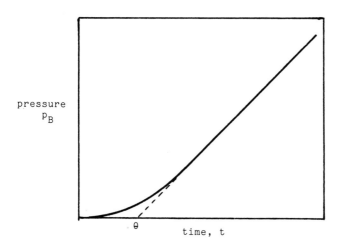

Figure 2   Plot of permeate pressure versus time

    Knowing the volume v_B and temperature T of B, the flux
J_t at any time t is given by

$$J_t = v_B/RT \ (dp_B/dt) \qquad\qquad (2)$$

After a long time the p_B versus t line becomes straight,
dp_B/dt is then constant and the steady flux is J_s. Given
the membrane dimensions and with p_B so small that Δp can be

replaced by p_A, the product DS can be evaluated from equation
(1). This product is usually called the permeability coefficient P.

During the early part of the experiment the time course of
the pressure in B can be inferred from a solution of Fick's
second law of diffusion with the appropriate boundary conditions.[8]
The diffusion coefficient can be obtained from such data in
several ways. The most familiar involves the so-called time
lag θ . This, as indicated in Figure 2, is the intercept made
with the time axis by back extrapolation of the straight region
of the p_B versus t curve. The relation between θ and D
was first given by Daynes[9] for the case where D is independent
of p_A. It is

$$D = h^2/6\theta \qquad (3)$$

The solubility then follows from the relation

$$S = P/D \qquad (4)$$

In more recent times many refinements in the handling of
permeation data have been introduced which permit S and D
to be evaluated when they are functions, of known form, of the
concentration of gas in the polymer.[8,10] Furthermore it is no
longer necessary to start with A and B evacuated. Solutions
of the diffusion equation exist for the relation between D and
the times of transition from one steady state to another and
these have permitted great economy in experimental time and
effort.[11]

With larger gas molecules D becomes smaller and S
usually larger. Even with a very thin film the time required
to reach steady permeation may well become relatively long.
Then it is easier to measure S directly in an absorption
or desorption experiment by using, instead of a film, a uniform
powder or microspheres of the polymer as the sample which reaches
sorption equilibrium more quickly. To find D one may use
another method of analyzing permeation data which depends on the

behaviour of p_B at very early times, long before steady
permeation is achieved.[12]

When the permeant is sorbed in an amount large enough to
swell the sample appreciably not only concentration dependences
of S and D but also time dependences have to be taken into
account. This is particularly likely with organic vapours and
gases, such as the lower hydrocarbons and carbon dioxide, at the
pressures likely to be useful in a separation process. Time-
dependent behaviour occurs because swelling causes stresses which
relax only slowly through changes in the internal conformations
of the polymer chains which are hindered by inter- and intra-
chain interactions.

Concentration and time variations of D cannot be disen-
tangled by combining time lag and steady flow data. Instead it
is usual to study the uptake or loss of vapour as a function of
time. A sheet of the polymer containing initially a uniform or
zero concentration of the penetrant is suspended from a balance
in an atmosphere of the vapour at some constant pressure and
temperature with which the sample is not in equilibrium. By
observing the changing mass of the sample until a new equilibrium
state is reached the sorption isotherm can be obtained. In most
cases the isotherm can be represented satisfactorily in the
Flory-Huggins form

$$\ln (p/p_o) = \ln (1-\phi_p) + \phi_p + \mathbf{X}\phi_p^2 \qquad (5)$$

Here p is the pressure in the vapour phase and p_o the
saturation pressure, ϕ_p the volume fraction of polymer in the
sample at equilibrium (i.e. $[1-\phi_p]$ is the volume fraction of
penetrant) and $\mathbf{X}$ the interaction parameter for the particular
polymer and vapour. Note that although equation (5) is more
complex than Henry's law it contains only a single parameter
to characterize the system.

Depending on the boundary conditions of the experiment and
on whether or not relaxation processes are important, the time

courses of absorption and desorption can take many forms. The full analysis of such data[8,13] is too complex to deal with here. Figure 3 illustrates the type of curve obtained in absorption when there is no time dependence. The mean diffusion coefficient $\overline{D}$ can be evaluated from the slope at short times of the plot shown because in this case the diffusion equation reduces to

$$\frac{q}{q_o} = \frac{4}{h}\left[\frac{\overline{D}t}{\pi}\right]^{\frac{1}{2}} \tag{6}$$

When D is independent of concentration it is simple to evaluate D from the time $t_{0.5}$ when the amount taken up q is just half the amount to be taken up at equilibrium q_o. Then

$$D = 0.04919\, h/t_{0.5}^{\frac{1}{2}} \tag{7}$$

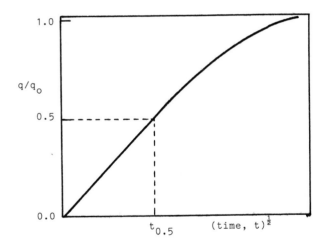

Figure 3   Plot showing the progress of sorption by a film

Usually when the best way of measuring D is to follow the kinetics of sorption, the concentration dependence of D cannot be neglected and may be determined by carrying out a suitable series of experiments.[13]

Under the conditions likely to be used in a vapour separation
process swelling stresses are likely to be fully relaxed but
the concentration dependences of the diffusion coefficients will
be important and complex since the concentration of each
component affects the permeation of all the others.

Although there are many other ways of studying diffusion
in polymers[7] the procedures summarized above, and their
variants, are the ones most suited to the polymers and penetrants
likely to figure in gas phase separation schemes. In the following sections the behaviour of characteristic types of systems
will be outlined drawing attention to the differences between
gases and vapours and to special features associated with
membranes where the polymers are in the rubbery and in the
glassy states. Finally the problems associated with mixed
permeants will be discussed.

Sorption and Diffusion of Gases in Rubbery Polymers
The distributions and heat motions of the chain segments
in polymers at temperatures well above the glass transition
resemble those in a normal liquid. Simple gases are not very
soluble in such polymers. At pressures up to several bar the
solution-diffusion model with constant S and D describes
these systems very well. Three variables are important: the
gas, the polymer and the temperature.

The solubilities of a series of gases in a given polymer
increase with increasing molecular weight and polarizability
of the gas molecules. There are several empirical, but fairly
successful, linear expressions correlating log S with the boiling
point, critical temperature or Lennard-Jones force constants
of the gas.[14] The values of S for a set of gases lie in the
same order in different polymers (Table 1) and decrease with
increasing polarity of the polymer at the same temperature
(Table 2). Data on many gases are available only in non-polar
elastomers and some correlations break down when they are tested
on more polar polymers. Thus polar gases such as SO_2 and CO_2
appear to increase in solubility with increasing polarity of the
polymer.

Table 1   Solubility coefficients S of gases in natural rubber
          and polyisobutene at 25°C

gas	$S \times 10^8 Pa$	
	natural rubber	polyisobutene
helium	11	11
hydrogen	39	34
nitrogen	52	52
oxygen	100	108
carbon dioxide	1010	699
acetylene	1640	638
ammonia	6990	1260
sulphur dioxide	23900	3650

Table 2   Solubilities S and diffusivities D of hydrogen and
          oxygen in several polymers at   43°C

polymer	hydrogen		oxygen	
	$S \times 10^8 Pa$	$10^{11} D/m^2 s^{-1}$	$S \times 10^8 Pa$	$10^{11} D/m^2 s^{-1}$
natural rubber	42	185	105	36
polyisobutene	41	31	109	2.4
polybutadiene	39	160	98	30
Neoprene	30	74	79	10
polyvinyl acetate	31	47	61	1.7

Solubility depends on temperature according to the Clausius
-Clapeyron equation

$$\Delta H = - R [d \ln S/d (1/T)] \qquad (8)$$

which permits the heat of sorption ΔH to be found.

It is seen in Table 3 that the heat varies only little
with the nature of the polymer but greatly, even as to sign, with
the nature of the gas. Sorption may be thought of as a combination of two events: an endothermic step of creating a molecular
size cavity in the polymer and an exothermic step of putting a
gas molecule into the cavity where it interacts with its

surroundings. For light, scarcely polarizable gas molecules
these interactions are weak and sorption is net endothermic;
for larger, more polarizable molecules the second heat out-
weighs the first and sorption is net exothermic.

Table 3 Heats of solution ΔH of gases in polymers

gas	$\Delta H/kJ\ mol^{-1}$		
	natural rubber	polyiso-butene	polyvinyl acetate
helium	7.5	7.5	8.8
hydrogen	3.3	2.5	10.3
oxygen	-3.3	-5.0	-4.6
carbon dioxide	-11.7	-8.8	-

Table 4 D and S for several gases in polyvinyl acetate
         at $40^{\circ}C$

gas	$10^{11}D/m^2s^{-1}$	$S \times 10^8 Pa$
helium	127	11.3
hydrogen	39.1	29.0
neon	26.2	12.4
oxygen	1.23	60.4
argon	0.397	84.6
krypton	0.0804	192.5

The range of values of D is far greater than that of S.
When one compares several gases in the same polymer (Table 4)
it is noted that an increase in S is accompanied usually by a
decrease in D while examining the same gas in several polymers
(Table 2) does not produce even a qualitative correlation
between D and S. Because the variations in D dominate, they
decide whether the permeability increases or decreases. Probably
the range of values covered by D would be much less if instead
of comparing data on different polymers at the same temperature
the comparison were made at equal distances above their glass
temperatures i.e. in states of comparable free volume. A further

complication in making such comparisons is that D is more
sensitive than S to changes in the shape of the permeant
molecules.[15] However this effect has not so far made possible
an efficient separation of isomers by permeation methods.

Perhaps the most useful empirical correlation is the
observation that the ratio of the permeabilities of a pair of
gases is surprisingly constant in a range of polymers over
which the absolute value of P may vary by a factor of a
million.[16] This observation has serious implications for those
attempting the development of more selective membrane polymers.

Diffusion coefficients are strongly temperature dependent
and follow the Arrhenius equation

$$D = D^o \exp(-E/RT) \tag{9}$$

fairly well; although accurate work shows the energy of
activation E decreasing slowly with increasing T.

Attempts have been made to correlate D, E and the pre-
exponential factor D^o with dimensions of the gas molecules
such as their collision diameters and least cross sections but
the significance of these correlations is doubtful. The most
convincing and informative is the linear correlation observed
between E and the squares of the collision diameter d of the
gas. Figure 4 shows just one example of this. From this
observation it has been inferred [17] that the mechanism by which
a gas molecule diffuses in a rubbery polymer is that a coop-
erative motion of about four roughly parallel polymer segments
opens a short passage in the polymer along which the diffusing
gas molecule jumps. This mechanism of activation and diffusion
has been discussed in some detail.[18] Table 5 shows that as the
gas molecules become larger E increases more and more slowly
towards an almost constant ceiling value. This suggests that
the mechanism of transport changes from the one outlined above
for small molecules to a process in which a gas molecule and a
polymer chain segment exchange places as a result of a local
kinetic disturbance.

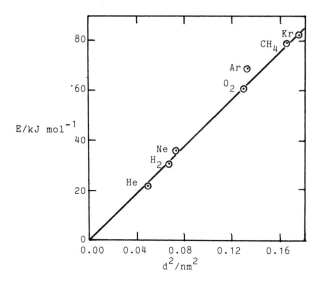

Figure 4   Energies of activation for diffusion versus the
           molecular diameter squared in polyvinyl acetate

Table 5   Energies of activation   E   of diffusion of gases
          in natural rubber

gas	E/kJ mol^{-1}
hydrogen	24.7
oxygen	31.4
nitrogen	36.4
carbon dioxide	37.2
methane	35.6
ethane	37.7
propane	37.7
n-butane	45.2
n-pentane	52.3

The Diffusion of Gases in Glassy Polymers
    Transport phenomena in the region of the glass transition
temperature T_g are complicated by relaxation processes
interacting with transport processes and by the fact that the

presence of dissolved gas may induce a glass transition to take place in the volume element around a gas molecule or around two gas molecules approaching one another in the polymer. These matters are not discussed here because they are unlikely to be of importance in separation processes, instead transport in the glassy state will be considered. A number of the polymers currently in use as membranes in separation processes are in the glassy state at process temperatures so this area of study is particularly relevant.

If D and S are measured as functions of temperature through a range from well above to well below T_g it is found that plots of log D and log S versus 1/T both change slope sharply at T_g although D and S have continuously varying values.

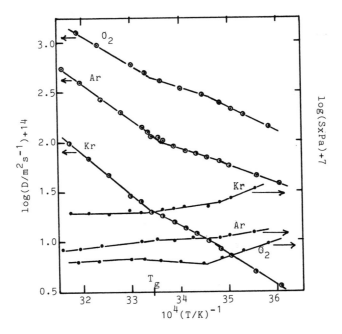

Figure 5   Logarithmic plots of D and S versus 1/T showing change of slope at T_g in polyvinyl acetate.

Figure 5 shows some data for gases diffusing in polyvinyl acetate.
Similar results have been noted with many polymers [19] except
that the second change of slope in Figure 5, that below T_g,
is not always seen and may have been caused by the polyvinyl
acetate used having some chain branching. It can be seen in
Table 6 that E is smaller below T_g than above, and if D^o
is used in conjunction with transition state theory to calculate
the entropy of activation ΔS^* for diffusion, this also is lower
below T_g than above.

Table 6   Energies E and entropies ΔS^* of activation for
          diffusion in polyvinyl acetate above and below T_g

gas	$E/kJ\ mol^{-1}$		$\Delta S^*/JK^{-1}mol^{-1}$	
	$T > T_g$	$T < T_g$	$T > T_g$	$T < T_g$
helium	22.4	17.4	-23.8	-20.9
hydrogen	31.4	21.6	-5.0	-19.2
neon	35.4	30.8	4.6	8.8
oxygen	60.6	46.4	59.4	31.8
argon	69.0	47.6	77.0	24.7
krypton	81.2	60.7	102.5	52.3

A consistent explanation of all these findings is that in
the glassy state, where chain motions are greatly restricted,
the activation steps for diffusion of gas molecules involves
only localized disturbances of short chain segments and,
consequently, diffusing molecules have shorter unit jump lengths
than in the rubbery state. The change over from short to longer
jumps occurs more or less sharply at T_g.

The fact that the heat of solution of gas below T_g is
always exothermic shows that the introduction of gas into the
polymer increases the total cohesive energy of the system.
Remembering also that the specific volume of a polymer in the
glassy state is greater than the value obtained by a linear
extrapolation from above T_g, suggests the presence of local
decrements in density sometimes called "holes" in the glassy state.

Further evidence for this notion comes from data on gas solubilities measured in different ways. Larger values are obtained below T_g from equilibrium sorption measurements than from the ratio P/D. This suggests that gas molecules may be strongly held in the "holes" and so be relatively immobilized and undetected in permeation experiments. If gas molecules absorbed in a glassy polymer can be thought of as distributed between two states viz. dissolved in the polymer with normal chain segment packing density and held by stronger cohesive forces in regions of low segment density, effects should appear in the permeation data that are not seen above T_g.

Dual Mode Sorption and Diffusion in Glassy Polymers

A theory, called the dual mode theory, has been developed to deal especially with gas permeation in glassy polymers.[21] In its original version, the theory regarded the gas molecules taken up as being present partly as a mobile population dissolved according to Henry's law in the bulk of the glassy polymer. This fraction of the gas diffuses down a concentration gradient in the normal way and with the kinetic characteristics outlined in the previous section. The remaining gas molecules are regarded as bound by a fixed number of adsorption sites in the polymer. This adsorbed fraction obeys the Langmuir isotherm and, in conformity with Langmuir's model, the adsorbed molecules are immobilized on their sites. Finally, it is assumed that locally the concentrations of bound and mobile molecules are at equilibrium.

In a permeation experiment, gas entering the membrane through the upstream face and diffusing through it will be partially trapped by the adsorption sites. The adsorption process prolongs the time required to set up steady flow and values of D in the bulk material calculated from the time lag by using equation (3) will be too small. Furthermore the ratio of P/D does not give a straightforward solubility S. A characteristic of such systems is that the total sorption isotherm can be expressed as the sum of a Henry's law solubility and a Langmuir adsorption term. Its shape therefore shows negative deviations from the linearity characteristic of Henry's law. The permeability of the membrane is found to decrease with increasing upstream gas pressure, the effect being more pronounced the more complex and more readily adsorbable the gas molecules.

The dual mode model has been given a detailed mathematical treatment by several workers.[22, 23, 24, 25] Although this cannot be dealt with here, two important results can be readily under-stood. The apparent solubility coefficient S_a (i.e. total sorbed concentration /pressure) is expressed by the sum of a Henry's law and a Langmuir adsorption term by

$$S_a = k_D + C_H b/ (1 + bp) \qquad (10)$$

where k_D is the solubility constant in the dense material, C_H the concentration in the "holes" when they are saturated with gas and b is the hole affinity constant. It can be seen from equation (10) that S_a decreases as p increases.

Assigning one diffusion coefficient D_D to the molecules dissolved in the dense material and another D_H, which is less than D_D and may be zero, to the adsorbed molecules, the perm-eability is given by

$$P = k_D D_D + D_H C_H b/ (1 + bp) \qquad (11)$$

where the first term $k_D D_D$ corresponds with DS in equation (4).

During the progressive development of this theory allowances have been introduced for the partial diffusivity D_H of the adsorbed molecules and for adsorption according to isotherms other than the Langmuir. Departures from local equilibrium between dissolved and adsorbed molecules have also been considered.

Many experimental studies on glassy polymers, particularly on packaging films but also on membrane polymers such as poly-sulphone, [26] appear to have confirmed its essential validity. Methods of data analysis have been devized to enable parameters such as the solubility and diffusivity in the dense material as well as the characteristics of the adsorption process to be evaluated.[27]

It should be understood that the sharp division of the gas
molecules sorbed in a glassy polmer into two fractions is necessarily an idealization of the real situation. It has been argued[28]
that carbon-13 rotating-frame relaxation rate measurements
indicate only a single population of gas molecules in a gas +
glassy polymer system and that interaction of gas and polymer
molecules results in the chain relaxation rates increasing
progressively with increasing gas concentration, i.e. external gas
pressure. On this evidence a new theory, the gas-polymer matrix
model, has been proposed which also predicts a solubility coefficient that decreases and a diffusion coefficient that increases
with increasing gas pressure in glassy polymers. The magnitudes
of these effects are such that the permeability is predicted to
decrease as pressure increases.

Sorption and Diffusion of Vapours in Polymers

In ordinary parlance the distinction between gases and
vapours is not sharp. Thus, for example, SO_2 and CO_2 are
usually thought of as gases although at ordinary temperatures
they are vapours. We are interested here also in other low
molecular weight vapours such as the lower hydrocarbons and
fluorocarbons. Even at low pressures significant amounts of such
substances are taken up by polymers. Their effect on the
dimensions and the properties of the polymer cannot be neglected.

Where the polymer is in the glassy state, either the experimental observations have been found to fit the dual mode theory[29]
but with a minimum in the permeability versus pressure curve,
that might be exploited in a separation process,[30] or the
concentration of the sorbed vapour has lowered T_g enough to
induce the glass transition in parts of the sample. Then
complex time-dependent phenomena are seen but they are not dealt
with here because separation is not likely to be carried out
under such unstable conditions.

At temperatures well above T_g, in the rubbery state, the
sorption isotherms fit equation (5) and are usually determined
by equilibrium uptake rather than by time-lag methods. The heat
of sorption is normally negative as the vapour essentially

condenses into the polymeric phase. Consequently the solubility
decreases with increasing temperature.

The diffusion coefficients of vapours are usually found to
increase markedly with increasing penetrant concentration (water
vapour is a special case which will be dealt with in a later
lecture in this conference). Any experiment gives a mean
coefficient $\bar{D}$ which is averaged across the concentration
interval $C_A - C_B$ of the experiment. The differential diffusion
coefficient $D(C)$ is found from

$$\bar{D} = \frac{1}{C_A - C_B} \int_{C_B}^{C_A} D(C) \, dC \tag{12}$$

Many experiments have to be carried out to determine the true
course of $D(C)$. Commonly its relation to concentration is
roughly exponential

$$D(C) = D_o \exp (\beta C) \tag{13}$$

where β is an empirical constant which decreases as temperature
increases.

A quantitative theory of this concentration dependence
has been developed from the idea that vapour diffusion is
governed by a volume of activation, a localized thermal density
fluctuation, which provides the space needed for a coordinated
regression of chain segemnts which enables a molecule from the
vapour to jump from one location in the polymer to another.[31]
The Arrhenius energy of activation is the excess energy associated
with an element of polymer containing a sufficient packet of extra
free volume relative to the average energy of such an element.

This so-called free volume theory of diffusion in polymers
has been progressively developed over a long period. The form
most convenient for practical application is that given by
Fujita[32] but a recent re-examination of the theory [33,34] has
enabled it to be significantly refined and extended beyond the
confines of vapour diffusion above T_g. Fujita's equation

expresses the observations on many vapour-polymer systems very
well and shows that over a wide concentration range equation (13)
is too simple. The theoretical expression for $D(C)$ is rather
complex but can be reduced to the form

$$D(C) = D_o \exp [AC/(B + C)] \qquad (14)$$

Here A and B are functions of a number of parameters which
may be determined from the densities and thermal expansion
coefficients of the polymer and liquid diffusate together with
the free volume fraction in the polymer at its T_g, which appears
to be almost a constant over a range of polymers, and a quantity
characteristic of the particular polymer and diffusate studied.
Comparison of equations (13) and (14) shows that the former
exaggerates the concentration dependence of D.

Permeation in more Complex Polymer Systems

In this brief survey it has been necessary to restrict the
coverage to gases and vapours transported in single amorphous
polymers. Polymers are however very versatile and varied
materials and many other systems are also important although less
well understood. There are many partly crystalline polymers in
which the crystallites act as impermeable inclusions. Their
effect on sorption and diffusion depends on their size and organ-
ization as well as on the crystalline/amorphous ratio. Dual mode
theory may be applied if the penetrant is adsorbed on the cryst-
allite faces or in the fringe regions around the crystallites
but a complex situation arises when the presence of the permeate
melts some of the less stable crystalline material.

Copolymers, including block and graft copolymers, as well as
miscible and immiscible blends of different polymers are extensively used to produce films. Some work has been done on their
permeabilities but little has emerged in the way of basic understanding because of the structural complexities especially when
the material is microheterogeneous.

Many polymer films are rendered anisotropic by uniaxial
or biaxial drawing. This greatly affects their transport

properties. Permeability studies of such oriented materials
have been carried out sporadically for more than thirty years but
an intensive programme of research in this area is needed now
because the effects produced might well be beneficially exploited
in membrane separation processes.

Permeation of Mixtures

There has been strangely little systematic work so far on
the permeation of mixtures of gases or vapours in polymers.
Three types of behaviour have been observed. In the simplest,
the flux of each gas is just what it would have been had the gas
been present alone at the same partial pressure. When one at
least of the gases is sufficiently sorbed to plasticize the
polymer its presence increases the flux of all other gases
present. Where dual mode sorption and diffusion are important,
two gases compete for the adsorption sites and the more strongly
adsorbed depresses the flux of the less strongly adsorbed
provided the adsorbed molecules have any mobility in the polymer.

Early studies of the permeation of gas mixtures dealt
only with permanent gases at low pressures and the polymers
usually were rubbers. Then the solubilities were low and the
chance of two gas molecules interacting significantly during
permeation was negligible. The behaviour of each gas was found
to be unaffected by the others and, if their individual permeabilities were measured, it was straightforward to predict the
fluxes and selectivities to be expected with a mixture of
permeants. Several of the membrane-gas separations in current
use fall into this category.

When one of the gases or vapours is more complex, such as
SO_2, CO_2, or the hydrocarbons, solubility is higher. At the
pressure needed for a practical separation process enough may
dissolve to plasticize the polymer if it is above its T_g. The
consequent rapid increase in D with concentration of permeant
has then to be taken into account. The extra free volume
imparted to the membrane by one or more components of the mixture
affects the permeabilities of all of them. The free volume
treatment of concentration-dependent diffusion has been extended

to mixtures by treating the free volumes contributed by each
component as being additive.[35,36] Qualitatively, the behaviour
of mixed permeants is characterized by larger fluxes and lower
separation factors than would have been predicted from the
permeation of each gas separately. This behaviour has to be
taken fully into account when selecting the membrane polymer and
the operating conditions for a separation process.

The greatest selectivities towards mixtures of permeating
gases are found in polar polymers having stiff chains. However,
their absolute permeabilities are very low. Fortunately such
polymers are suitable to be formed into ultra-thin films either
as asymmetric membranes or in thin film composites. The thinness
helps to compensate for their low inherent permeabilities.

Such polymers are hard and normally in the glassy state
at the temperature of use. The consequences of dual mode sorption and diffusion have therefore to be taken into account. In
the dense regions, where the normal solution-diffusion mechanism
operates, the solubilities are low and the fluxes of the various
components are essentially independent and non-interacting. In
the transport pathway that depends on the adsorption and flow of
molecules in the regions of low density and high energy in the
polymeric glass, the behaviour of the gases is not at all
independent. The components have to compete for the adsorption
sites, as mentioned above. The observed behaviour is complex
and individual to each system. It is only now beginning to be
studied systematically.[37,38] It appears that the fluxes of both
components of the gas mixture are lower than the permeabilities
of each measured alone would lead one to expect but the effects
are unequal and usually lead to an enhanced separation factor
which is capable of practical exploitation. It may be noted that
in the opinion of the author the dual mode theory and the gas-
polymer matrix model lead to different predictions about the
behaviour of mixtures of permeants. A critical test of these
rival theories might therefore be devised from the data now
appearing on systems of mixed permeants.

24   *Fourth BOC Priestley Conference*

References

1 J.K. Mitchell, Royal Inst. J., 1831,2, 101,307.

2 T. Graham, Phil. Mag., 1866,32,401.

3 S. Weller and W.A. Steiner, J.Appl.Phys., 1950,21,279;
   Chem.Eng.Prog., 1950,46,585.

4 V.T. Stannett,W.Koros,D.R. Paul,H.K. Lonsdale and R.W.Baker,
   Adv. Polymer Sci.,1979,32,69.

5 P.J. Flory, "Principles of Polymer Chemistry",Cornell
   University Press,Ithaca,N.Y.,1953,Chapter 12.

6 R.E. Kesting,"Synthetic Polymeric Membranes",Wiley,New York,
   1985, Chapter 7.

7 R.M. Felder and G.S. Huvard,"Methods of Experimental Physics",
   Academic Press,New York,1980,Vol.16,Chapter 17.

8 J. Crank,"Mathematics of Diffusion",Oxford University Press,
   London,2nd Ed.,1975.

9 H.A. Daynes, Proc.Royal Soc.A, 1920,97,286.

10 H.L. Frisch, J.Phys.Chem.,1957,61,93;1958,62,401.

11 S.G. Amarantos and J.H. Petropoulos,J.Membrane Sci.,1979,
   5,375.

12 P. Meares, J.Appl.Polym.Sci., 1965,9,917.

13 G.S. Park in J.Crank and G.S.Park (eds),"Diffusion in
   Polymers", Academic Press,London,1968,Chapter 5.

14 G.J. Van Amerongen, Rubber Chem.Tech.,1964,37,1065.

15 A.Aitken and R.M. Barrer,Trans.Faraday Soc.,1955,51,116.

16 S.M. Allen,M.Fujii,V.Stannett,H.B.Hopfenberg andJ.L.Williams,
   J.Membrane Sci.,1977,2,153.

17 P.Meares, J.Amer.Chem.Soc., 1954,76,3415.

18 C.A.Kumins and T.K.Kwei in J.Crank and G.S.Park (eds.),
   "Diffusion in Polymers", Academic Press,London,1968,Ch.4.

19 V. Stannett, J.Membrane Sci.,1978,3,97.

20 P. Meares, Trans.Faraday Soc.,1958,54,40.

21 W.R.Vieth,J.M.Howell and J.H.Hsieh,J.Membrane Sci.,1976,1,177.

22 D.R.Paul andJ.W.Koros,J.Polym.Sci.Polym.Phys.,1976,14,675.

23 J.H. Petropoulos,J.Polym.Sci.,A-2,1970,8,1797.

24 Z. Grzywna and J.Podkowka, J.Membrane Sci.,1981,8,23.

25 R.M.Barrer, J.Membrane Sci., 1984,18,25.

26 A.J.Erb and D.R.Paul, J.Membrane Sci., 1981,8,11.

27 D.R.Paul, Ber.Bunsenges.Phys.Chem., 1979,83,294.

28 D.Raucher and M.D.Sefcik, "Industrial Gas Separations",ACS
    Symp.Ser.223,American Chemical Society,Washington,D.C.,
    1983,Chapters 5 and 6.

29 S.R. Mauze and S.A. Stern,J.Membrane Sci., 1984,18,99.

30 S.A.Stern and V.Saxena, J.Membrane Sci.,1980,7,47.

31 P. Meares,J.Polym.Sci., 1958,27,391.

32 H.Fujita,Adv.Polymer Sci., 1961,3,1.

33 J.S.Vrentas andJ.L.Duda,J.Appl.Polym.Sci., 1978,22,2325.

34 J.S.Vrentas,J.L.Duda and H.-C.Ling,J.Polym.Sci.Polym.Phys.,
        1985,23,275,289.

35 S.-M.Fang,S.A.Stern and H.L.Frisch,Chem.Eng.Sci.,1975,30,773.

36 M.Fels,Amer.Inst.Chem.Eng.Symp.Ser.,1972,68(120),49.

37 R.T.Chern,W.J.Koros,E.S.Sanders,S.H.Chen and H.B.Hopfenberg,
    "Industrial Gas Separations",ACS Symp.Ser.223,American
    Chemical Society,Washington,D.C.,1983,Chapter 3.

38 E.S.Sanders,W.J.Koros,H.B.Hopfenberg and V.T.Stannett,
        J.Membrane Sci., 1983,13,161.

Conditions Which Need to be Fulfilled by Membrane Systems in Order to Compete with Existing Methods of Gas Separation

By W. Baldus and D. Tillman*

LINDE AG, WERKSGRUPPE TVT MÜNCHEN, 8023 HOLLRIEGELSKREUTH, BEI
MÜNCHEN, FRG

1   The separation of gas mixtures is carried out for the
purpose of obtaining one or more of the components in a
specified purity. This technique plays an important role
in an immense number of large-scale processes. The
separation is done by various methods. These methods are
based on different physical and/or chemical properties
of the materials, e. g. the different boiling points of
the components, special adsorption or absorption
properties, a different diffusion behaviour, different
equilibria and many other things. Consequently, the
processes work e.g. with different temperatures or
different adsorption materials or scrubbing systems.
These processes compete with each other and the nearly
unlimited variety of their application makes it
difficult to decide which process is the most economical
one for a certain purpose. This question, however, is
naturally of greatest importance in the technical field.

In the last 10 years a new technique for the separation
of gases has emerged, namely the permeation through
membranes. We should like in the following to ask the
question: in which cases can the membrane technology be
applied with good prospects and in which cases are there
other more favourable processes. It will also be shown
that a combination of a membrane step with other process
steps can lead to an economical process. We shall try
with the aid of examples to clarify the subject.

1.1   The various competitive processes are based on specific
physical properties. Characteristic for:

- the condensation and distillation processes are
  the K-(equilibrium) values

- the scrubbing processes are
  the solubility values and the chemical
  binding forces

- the pressure and temperature swing adsorption
  processes are the adsorption isotherms

- the membrane processes are
  the selectivity and the permeability.

This, however, is only a very rough and simplified
survey. The practical conditions are much more
complicated. Nevertheless, one can say:

The characteristic properties given above are valid for
a certain gas mixture and also depend on different
parameters. These parameters can be split into two
groups:

- those which can be varied by the engineer

- and those which are determined by the problem of
  separation itself and which cannot be varied by the
  engineer, e.g. binding forces, permeability.

1.2   The final decision which of the competitive processes is
the best can, of course, only be made by an economical
comparison, i.e. which process has the lowest total cost
for a given separation problem. This, however, requires
that the engineer has chosen the optimal parameters -
temperature, pressure etc. - for each of the competitive
processes. The ecomomical parameters such as

- operating time
- depreciation rate (straight-line method)
- maintenance and labour prices
- energy prices, including electrical power, cooling
  water etc.

have to be identical for all processes.

1.3   In a membrane process, the decisive parameter is the
        selectivity. We define selectivity as follows:

$$S = \frac{\text{Permeation Rate of faster comp.}}{\text{Permeation Rate of slower comp.}}$$

The permeation rates should be measured on a
commercially available module. Therefore, it should be
kept in mind that in the following the selectivity
stands for an "Overall Separation Factor" which includes
all imperfections of the module, e.g. uneven flow
distribution etc. Nevertheless, the selectivity mainly
depends on the membrane material. For simplification, we
shall assume that the permeabilities do not depend on
pressure and concentration.

In theory, a process with a membrane of any desired
selectivity may be calculated. This is done in such a
way that the permeation rate of the faster component is
increased and that of the slower component is kept
constant. For reasons of comparison, the selectivity of
the membrane may be chosen so that the annual total
costs for the membrane process are equal to the costs of
a competing process.

We shall now demonstrate with three examples which
conditions - especially for the selectivity - need to be
fulfilled by membrane systems.

2   Purge Gas Separation of an Ammonia Plant

        Fig. 1
        shows the principle. It is based on the steam reforming
        process. In our case, attention is paid only to the
        purge gas separation after the ammonia synthesis. The
        purge gas is withdrawn from the ammonia synthesis loop
        in order to control the level of inert gases such as
        argon (from the air) and methane (from a methanator not
        shown in Fig. 1). The purge gas quantity depends on the
        enrichment factor in the loop and is of the order of 320
        kmols/hr for a 1000 t/d ammonia plant. It consists not

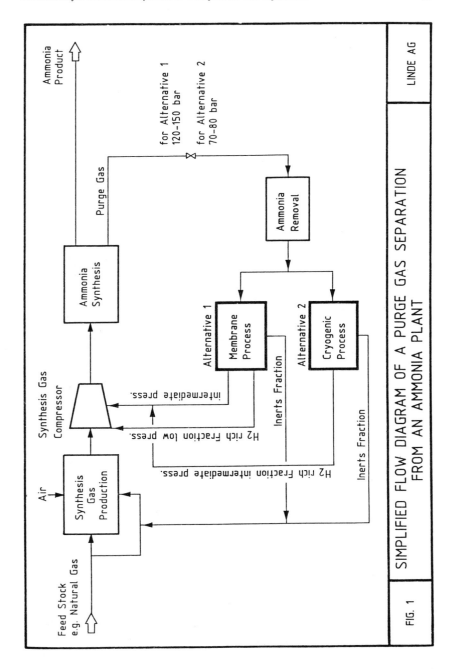

FIG. 1

SIMPLIFIED FLOW DIAGRAM OF A PURGE GAS SEPARATION FROM AN AMMONIA PLANT

LINDE AG

only of argon and methane, but also of nitrogen, ammonia
and especially of hydrogen. This hydrogen is used by
feeding it back into the synthesis loop. For the
separation of the hydrogen various processes are
available. Two are investigated in more detail: the
cryogenic and the membrane process.

2.1   Fig. 2

The membrane process (Alternative 1) consists of a
number of membrane modules, either the spiral wound type
or the hollow fibre type. These modules are arranged in
two groups. In the first one the high partial pressure
of the hydrogen in the purge gas (approx. 90 bars)
allows a high permeate pressure as well. This permeate
from the first group of the membrane modules is a
hydrogen rich fraction. It is fed to an intermediate
stage of the synthesis gas compressor. The retentate of
the first group enters the second group of membrane
modules. For these a lower permeate pressure is provided
in order to obtain a high hydrogen recovery rate
(approx. 90 - 95 %). The permeate from the second group,
also a hydrogen rich fraction, is fed to the first stage
of the synthesis gas compressor. The remaining gas
contains 70 to 80 % of the inerts (argon and methane).
This inerts fraction is used as fuel for the steam
reformer.

2.2   The cryogenic process (Alternative 2) works as follows:
after the NH_3 removal system traces of water and
ammonia have to be removed from the purge gas by an
adsorber. In the heat exchanger the gas is cooled and
most of the inerts (approx. 80 - 95 %) and part of the
nitrogen is liquefied against the cold hydrogen and
evaporating inerts fraction. In this case, a plate fin
heat exchanger is used. The purge gas is therefore
depressurized upstream the NH_3 removal unit to the
maximum allowable pressure (approx. 70 - 80 bars). The
liquid from the separator is expanded to fuel gas system
pressure. The hydrogen rich fraction is fed to the
intermediate stage of the synthesis gas compressor.

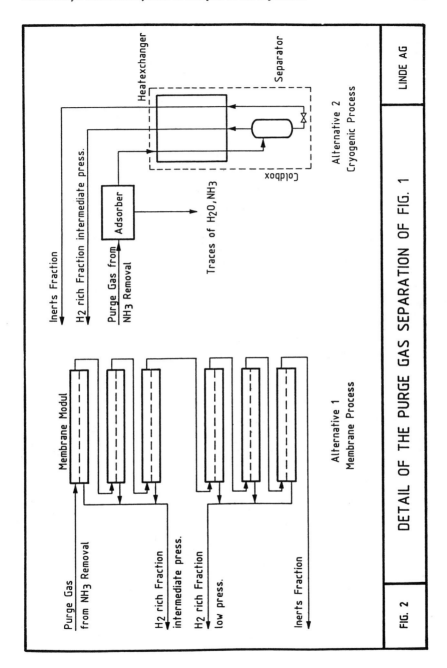

DETAIL OF THE PURGE GAS SEPARATION OF FIG. 1

FIG. 2

LINDE AG

2.3   <u>Fig. 3</u>

It shows the inerts removal rate versus the hydrogen
recovery rate of four processes:

1.   the membrane process

2.   the cryogenic process

(These two processes have just been described)

3.   the high pressure cryogenic process, which is very
     similar to Alternative 2, only the operating
     pressure is nearly at the same level as the
     synthesis loop pressure.

4.   the pressure swing adsorption process – it works at
     a lower pressure level but high hydrogen purities
     are obtained.

Process 1 and Process 2 in this list have proved to be
the most economical for purge gas recovery based on the
steam reforming process previously shown.

The membrane process – in this case –has the following
benefits in comparison to the cryogenic process:
–   higher hydrogen recovery rate
–   lower investment cost, especially for smaller plants
     and in those cases where the synthesis gas
     compressor has enough spare capacity
–   shorter delivery time
–   easily operated (start-up at once).

The calculations for this example are based on
selectivities of $H_2/Ar \approx 40$ and $H_2/CH4 \approx 50$. Membranes
with such selectivities are available and compete with
the cryogenic process.

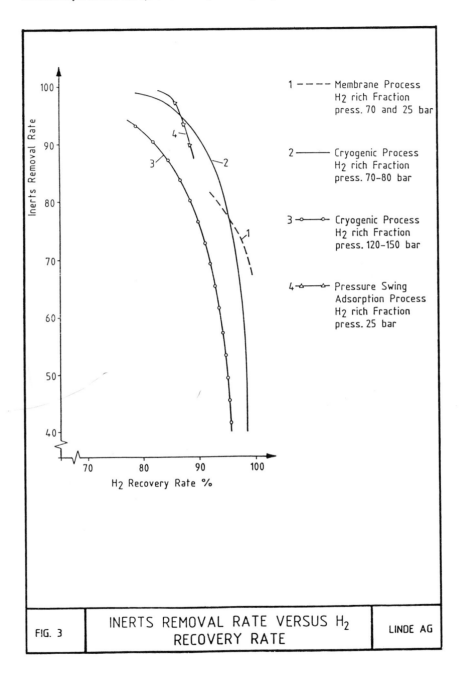

| FIG. 3 | INERTS REMOVAL RATE VERSUS H_2 RECOVERY RATE | LINDE AG |

The cryogenic process has the following advantages:
- the total hydrogen rich fraction is recovered at an intermediate pressure (70 - 80 bars)
- the inerts removal rate is high; therefore, the total amount of purge gas is low
- with the hydrogen a considerable amount of nitrogen is recovered, this may be important especially in those cases, where no spare capacity in the air compressor is available.

These advantages - each of them leads to lower energy costs - are normally compensated by the higher investment cost for the cryogenic process.

In order to decide which technology should be applied - croygenic or membrane - for the recovery of hydrogen from an ammonia plant purge gas, not only the lowest total costs of the process (see 1.2) must be compared, but also the operation and economic conditions of the ammonia plant itself have to be considered, e.g. the on-stream time of the ammonia plant.

3   Natural Gas Sweetening

Fig. 4

This example is a separation problem which is important in enhanced oil recovery. The water-saturated natural gas consists of 64 mol.% CO_2, 29 mol.% hydrocarbons and 7 mol.% nitrogen. A sales gas with max. 3 mol.% CO_2 and a CO_2 product containing a min. of 97 mol.% CO_2 have to be produced. Various processes, e.g. cryogenic, scrubbing and membrane processes have been investigated. Here, however, two processes, both applying membrane technology have been compared.

3.1   The first, with membranes only, works as follows: the natural gas, at a pressure of approx. 45 bars enters the first membrane unit. CO_2 of the required purity is withdrawn as permeate. The retentate from membrane one flows through a second and third membrane unit in

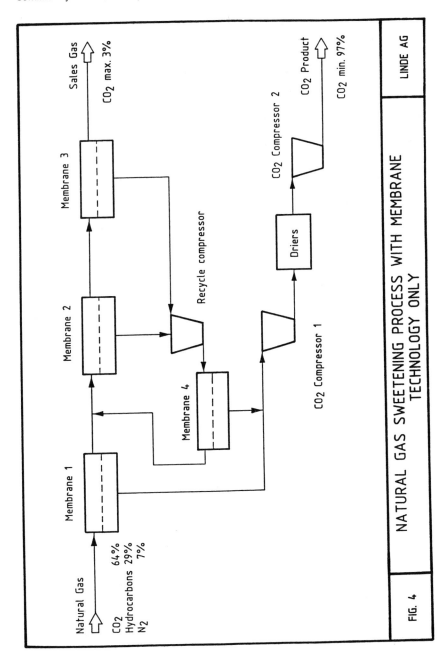

NATURAL GAS SWEETENING PROCESS WITH MEMBRANE TECHNOLOGY ONLY

FIG. 4

LINDE AG

series. There, on the permeate side, CO2 rich fractions
are obtained at two different pressures, 4.8 bars and
1.6 bars. The retentate from unit three represents the
sales gas with max. 3 mol.% CO2 and is available at
approx. 44 bars. The two permeate streams from membrane
two and three are recompressed to the feed pressure and
fed to the fourth membrane unit. The retentate from this
unit is mixed with the retentate from the first one. The
permeate, containing a minimum of 97mol. % CO2, i.e.
product quality, is mixed with the permeate from the
first membrane unit. It is compressed to the required
CO2 product pressure. Because of the fact that water
permeates faster in comparison to the other components
of the feed gas, nearly all of it is found in the CO2
rich fractions. A drier is therefore installed between
the CO2 compressor one and the CO2 compressor two.
Depending on the distribution of the molecular weight of
the hydrocarbons in the feed gas, a dew point control
system may be installed between the membrane units one
and two or three.

3.2   Fig. 5

The second process is a combination of cryogenic and
membrane technology. The feed gas, dried at the front
end of the process, is mixed with the recompressed CO2
rich recycle gas from the permeate side of the membrane
unit. The mixture is cooled down to the column inlet
temperature in the heat exchanger counter currently
against the overhead product of the column and
evaporating refrigerant. In the column, CO2 with the
specified product purity is separated from the lighter
components and withdrawn from the bottom. It is pumped
to the required CO2 product pressure. The top product,
containing approx. 36 mol.% CO2, is warmed up and fed to
the membrane. The retentate stream, withdrawn from the
membrane unit has sales gas quality at the required
pressure.

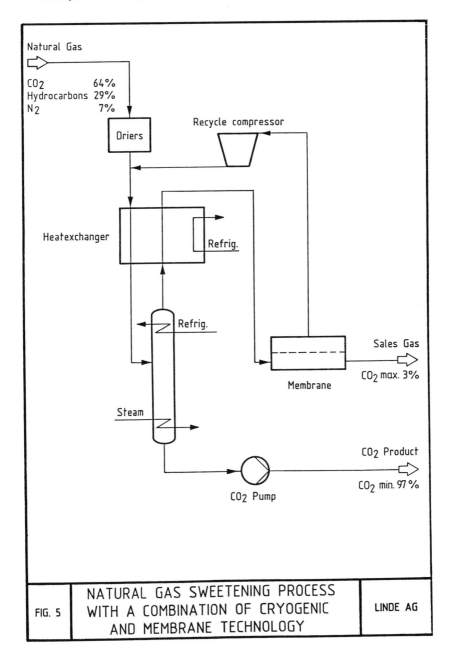

FIG. 5 — NATURAL GAS SWEETENING PROCESS WITH A COMBINATION OF CRYOGENIC AND MEMBRANE TECHNOLOGY — LINDE AG

3.3   An ecomomical comparison of the processes 3.1 and 3.2
       (based on 2700 kmols/h feed gas) shows a considerable
       advantage of the combined process 3.2 against the pure
       membrane process 3.1.

       In this case the calculations were based on a
       selectivity $CO_2/CH_4 \approx 20$. The question was: how must the
       selectivity be increased, so that the membrane process
       3.1 has the same efficiency as the combined process 3.2?
       The answer was: the selectivity must be increased up to
       the region of 50.

       This example shows that combinations of membrane
       technology and conventional separation steps should be
       taken into consideration before a process is chosen for
       a given separation problem. This is the case especially
       for larger plants where one of the characteristics of
       the membranes – the linear function of price and plant
       size – turns into a disadvantage.

4.   Production of Nitrogen from Air

       This example compares the production of nitrogen by a
       pressure swing adsorption process (PSA) and a membrane
       process. The nitrogen may be used as inert gas where
       small and easily operated plants are required, e.g. on
       offshore platforms and tank ships.

4.2   Fig. 6
       The PSA process consists of an air compressor, two
       adsorber vessels – where molecular sieve adsorbs water,
       carbon dioxide and oxygen – and product a buffer vessel,
       which balances fluctuations in pressure and
       concentration, caused by the alternately operating
       adsorbers.

4.2   The membrane process is characterized by the same
       simplicity as the PSA process. The air is also
       compressed first and then fed to a number of membrane
       elements which are arranged in series in respect of the

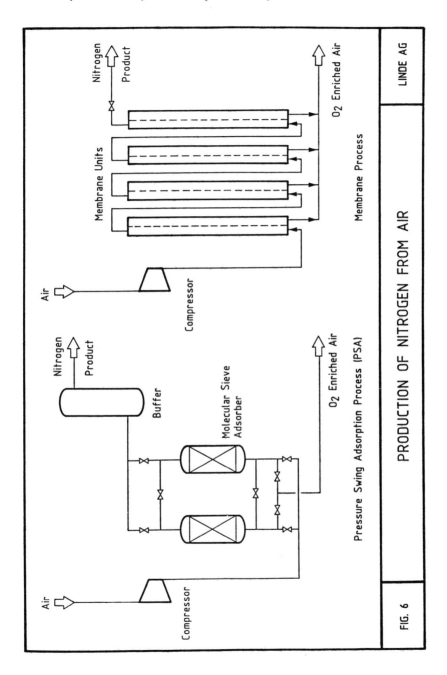

LINDE AG

PRODUCTION OF NITROGEN FROM AIR

FIG. 6

retentate and parallel in respect of the permeate. The
retentate represents the nitrogen product and is
obtained slightly under feed pressure. The permeate -
oxygen-enriched air - is released to the atmosphere. In
both cases the air compressor can be omitted if
compressed air from a system - e.g. instrument air
system - is available.

4.3   Because of the great simplicity, the investment and all
related costs for both processes are low and rather
similar. Therefore, the operating costs, especially
energy costs, are decisive.

Fig. 7
For a required purity of nitrogen the process with the
higher nitrogen recovery rate causes the lower specific
energy consumption. Fig. 7 shows the nitrogen recovery
rate versus the purity of the nitrogen product of the
PSA process and membrane process with different feed
pressures and separation factors (selectivities).

For the above described membrane arrangement and the
separation factor $O_2/N_2 \approx 4.5$ the nitrogen recovery rate
is always lower than that of the PSA process, even for
higher feed pressure. But an expected increase in
selectivity, however, will make the membrane process
competitive.

5.   Final Remarks

The discussed examples show the difficulty in deciding a
priori which technology - membrane or conventional -
should be applied for a given separation problem. In
each case the technical and economic conditions have
to be examined carefully. In addition, combinations of
membrane and conventional technology should be studied.
But some simple heuristic rules may be given:

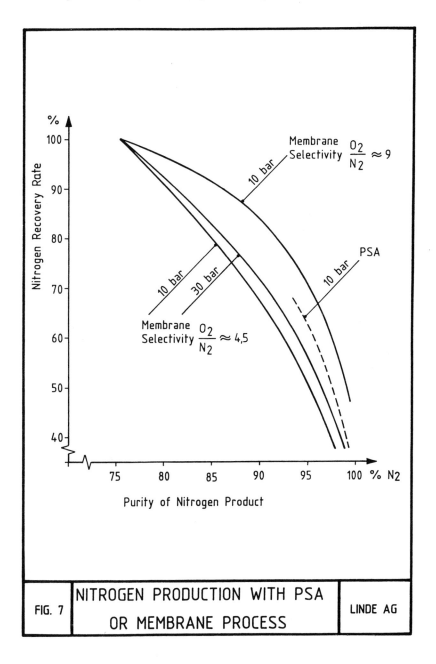

FIG. 7 | NITROGEN PRODUCTION WITH PSA OR MEMBRANE PROCESS | LINDE AG

The application of membranes for a gas separation
problem is favourable

- when moderate purity and recovery rates are
  sufficient
- when the components, which have to be separated are
  contained in the feedgas in considerable amounts and
  not only in traces
- when the feed gas is available at the necessary
  pressure or the retentate is required at high
  pressure
- when the feedgas contains no components which alter
  the properties of the membrane (or even destroy it).
- when a membrane with sufficient selectivity for the
  given separation problem is available

The membrane technology will certainly play an
increasingly important role in the gas separation field.
Especially combinations of membranes with conventional
process steps such as scrubbing, cryogenic separation or
adsorption may improve the economics of a number of gas
separation processes.

Some Technical and Economical Aspects of Gas Separation by Means of Membranes

By U. Werner

DEPARTMENT OF CHEMICAL ENGINEERING, UNIVERSITY OF DORTMUND, PO BOX
500 500, 4600 DORTMUND 50, FRG

It was Graham, a Scotsman, who was the first to recognize exactly
120 years ago /1/ that a selective gas separation can be performed
by means of different permeation velocities of various gases
through a membrane. It was not until the early 40s this century
that gas separation through membranes became technically interesting in order to enrich uranium 235 from about .71% of its natural
recources to about 3% by weight. Cost did not play any important
part for this special extremely energy consuming process. At that
time, there were no alternative enrichment processes available.
In the 50s and 60s, experiments were run at various places on a
minor scale on membrane units for gas separation. In principle,
however, there was a lack of membranes with adequately high permeability and separation efficiency in order to allow for a commercial
use. The solution to this problem apparently became available by
the invention of an asymmetrically structured membrane, the
manufacture of which was described by Loeb and Sourirajan in 1960
/2/. It consists of a very thin, nonporous, selective skin layer
resting on a highly-porous nonselective support layer. Finally, it
was the American Monsanto Company who, after development and in-house
tests, made the first breakthrough on the market in the 70s by
its composite membrane in the " Prism - Module " in H_2- Separation
from hydrogenation- and NH_3- production processes, respectively
/3/.
Current publications, patent applications and market launchings
indicate tumultuous developments which, originating from the USA
and Japan, apparently focus on the following major fields of application /4,5 6,7/ :

- H_2 -separation from process gases (NH_3, CH_4, CO etc.)
- O_2 -enrichment from air (oxidation processes)
- N_2 -enrichment from air (blanketing)
- CH_4 -enrichment from natural or waste gas
- CO_2 -enrichment from natural or waste gas
- H_2O -separation from air or industrial gases (drying).

Now , I should like to try to briefly outline to you the essential principles of gas separation through homogeneous membranes, to discuss some rather technically relevant criteria on the design and assessment to compare the membrane technique with alternative unit operations of chemical process engineering.

Gas Transport in Membranes

Apart from other feasible approaches, the so-called solution-diffusion model bears a technically adequate possibility for describing the processes occuring when various gases pass through a homogeneous membrane. It combines the well-known 1st Fick's law of diffusion

$$J_i = n_i/A = - D_i (dC_i/ds) \qquad (1)$$

with Henry's law on the solubility of a gaseous constituent in the matrix of the membrane material

$$C_i = H_i * p_i \qquad (2)$$

where J_i represents the molar mass-flow n_i (mol/time) of component i related to the area A. D_i is the coefficient of diffusion, C_i the local concentration, p_i the local partial pressure and H_i the solubility coefficient of component i in the membrane material. s is the linear coordinate according to the thickness of the membrane.

If Eq.(2) is substituted for Eq.(1) and integrated with respect to the thickness S of the membrane, it follows

$$J_i = (Cp_i/S)*(p_{ih} - p_{il}) \qquad (3)$$

where

$$Cp_i = D_i * H_i \qquad (4)$$

represent the permeability of gas component i, p_{ih} and p_{il} the partial pressures of component i on the high-h- and low-l-pressure-side of the membrane. For several gas components, such a flux-equation (3) must be given for each single component i,j,... The capability of the membrane to separate the components i and j of a binary gas mixture can be expressed by the selectivity

$$\alpha_{ij} = Cp_i/Cp_j \qquad (5)$$

which relates the ideal permeabilities of both gas components.

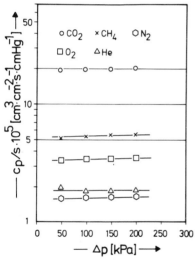

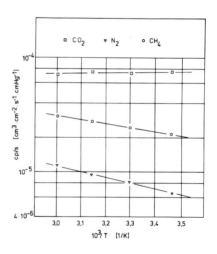

Figure 1 :
Permeability of Polydimethylsi-
loxan as a function of pressure
difference across membrane

Figure 2 :
Permeability of Polydimethylsi-
loxan as a function of temperature

(The indeed possible separation of gases by means of porous membranes made of glass, ceramic, metal, polymers etc. will not be
the subject of this paper. The selectivities are too small to be
economically usable for "normal" gases and temperatures /7a/.)

The permeability, Eq.(4), of a membrane for a gas component pre-
sents itself as a product of solubility and diffusion coefficient.
Whatever varies one of these two values does also vary the permeability. Primarily, it concerns the thermodynamical variables of
state, viz. pressure and temperature. Only for those gases the
critical point of which is in the normal temperature range (CO_2,
H_2S, H_2O, higher HC, etc.) sometimes results an appreciable dependence of the permeability on pressure itself, which needs to be
considered, Fig.(1). For this case also see pervaporation and vapor
separation /8/. For the majority of technically relevant gas mixtures the temperature dependence of permeability is of prime importance. Over a wide range of temperature, the dependence follows the
Arrhenius' plot

$$Cp_i(T) = K_i*exp(-E_i/RT) \qquad\qquad (6)$$

K_i as proportionality coefficient and E_i as activation energy
should be regarded as specific for the correlation between gas
component i and polymer, that means solubility and diffusivity, see
Fig.(2). But consequently the selectivity also becomes a function
of temperature

$$\alpha_{ij}(T) = (K_i/K_j)*exp(-(E_i - E_j)/RT) \qquad\qquad (7)$$

It becomes evident that the permeability increases with rising
temperature, while the selectivity decreases in most cases (i is
the faster permeating component).

If a membrane possesses an adequately high selectivity and mechanical strength, then an appreciable flexibility in the application
of the membrane process can be obtained by keeping the temperature
at certain levels /4/.

Table 1 lists a lot of permeability and selectivity data for
various polymers and pure gases as gathered from literature
/4,9,10/. It clearly indicates what membrane material should be
specifically suitable for what application.

Especially for the separation of O_2 and N_2 some data are compiled
which can be graphically linked in Fig. 3. It seems important that
in general the same trend results for the majority of gas mixtures
to be separated :

   The greater the selectivity and, consequently, the purity of a
   permeate that is desired to be attained for a technical separa-

Table 1: Permeabilities c_p and Ideal Selectivities α_{ij}

c_p [Ba]	PSU (Permeo)	CA (Envir.)	PI (Ube)	PMP (Dow)	PPO (Permeo)	SAN	PP	PMMA	PE	PC	EC	PS	PDMS
H_2	13	12	9	136	22				8.7	12			649
N_2						0.046	0.3	0.23	1	0.3	2	7.8	281
O_2	1	1	0.5	32	2.3	0.35	1.6	1.2	5	1.4	8.7	24	604
CO_2	6	6		93	9	1.1	6.8		13.2	8	47		3230
CH_4									3				1070
α_{ij}													
H_2/N_2	72	70	~200	17	56				8.7	40			2.3
H_2/CH_4	60	60	~200		55				2.9				0.6
CO_2/CH_4	25	30		4	22				4.4				3
O_2/N_2	6	5.5	11		5.9	7.6	5.4	5.2	5	4.7	4.3	3.1	2.1

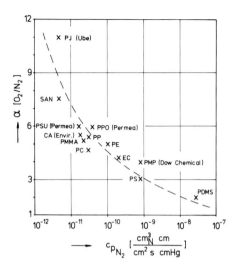

Figure 3 :
Selectivity α_{ij} of different
Membrane-materials as a Func-
tion of Permeability of N_2;
System O_2/N_2;
For data see Table 1.

tion process, the lesser the permeability, that means the permeate flux one will have to put up with.
This does not only apply when comparing the data of entirely different polymer grades, but also if proceeding from a basic substance (for inst. polyimid) and modifying it by varying the
polymerization or manufacturing conditions of the membrane / 5 /.
How strongly general thoughts and conclusions in terms of engineering may be taken into account in comparing and, ultimately,
selecting a membrane system is shown by the analysis, though com-
pany-biased but nevertheless apt and right, of recently launched
developments by Kurz et al./ 4 /.
In order to guarantee a certain flow through a membrane, the polymer material as per Eq.(3)+(5) should possess the least possible
thickness of the separating layer, a high permeability and a high
selectivity. The mechanical properties should feature a high stability against deformation (extension, compaction, collapsing)
over a large range of temperature. The membrane material should be
as chemically resistive as possible to traces occurring in technically processed gases (H_2S, higher hydrocarbons, organic solvents
etc.). Besides, the polymer material should be as cheap as possible
and be easy to process to a usable membrane, that means at low
cost. A requirement which should not be underrated is that the
membrane should be easy to handle and, last but not least

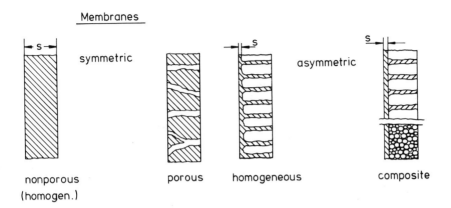

Figure 4 : Different Types of Membranes

be easy to connect to the supporting and casing assemblies by
means of simple techniques, e.g. by pressing, sticking, welding
etc. Most of these requirements can be met and satisfied by the
production of asymmetrical or composite membranes, Fig. 4, while
others are difficult to meet or, by nature, (linkage of permeability and selectivity) hardly or not at all possible. For manufacturing processes see /11,12/.

Plant and Equipment

Membranes of various structure and composition can be manufactured
as flat films or tubular as tubes or capillaries. They are processed in a lot of variants in plate-and-frame or spiral-wound modules, tube or capillary modules which represent the actual technically applicable separation stages, Fig. 5. Besides, the module
conception, above all the combining and sealing techniques, frequently incorporate as much know-how as the manufacture of the membrane itself. They differ mainly in regard to the packing density
(m^2 membrane area/construction volume), but also in respect of the
pressure drop behaviour on high- and low-pressure side which may
noticeably diminish the separation efficiency of such a system
/4,8/. Depending on the separation problem, single or many modules
are wired in the process in varying forms, Fig. 6. The selection of
a specific circuit form results of tasks, that means which types of
gases are to be separated, which fluxes occur at what pressure, how

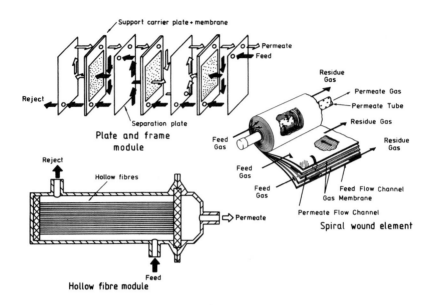

Figure 5 : Commonly used Module - Constructions

high the product concentration shall be raised and what recoveries
are demanded.

In order to outline the crucial design parameter, let us have a
look at the processes taking place in a fictitious membrane separation appliance, Fig. 7.

On one side of the membrane, the feed n_F being at a higher pressure
p_h is conveyed, in the simplest case of a binary gas mixture (component A and B) with the molar fraction x_F for the faster
permeating component A at the inlet to the module. On the other
side, the permeate n_p flows at a lower pressure p_l with the concentration rising to the molar fraction y_p of component A at the end
of the module.

The faster permeating component A passes more and more to the low
pressure side, that means it declines in concentration on the high
pressure side, and the slower permeating component B enriches in
the reject n_R to attain the outlet concentration x_R of   component
A. Concurrently, the driving partial-pressure-difference of the
faster component decreases and that of the slower component
increases.

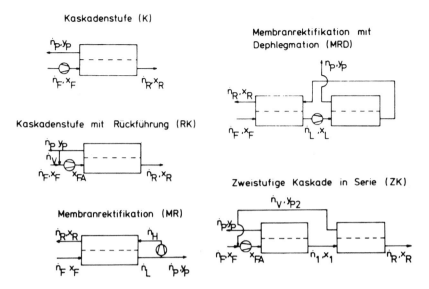

Kaskadenstufe (K)

Membranrektifikation mit Dephlegmation (MRD)

Kaskadenstufe mit Rückführung (RK)

Zweistufige Kaskade in Serie (ZK)

Membranrektifikation (MR)

Figure 6 : Flow- Circuits in Single- or Multi- Stage- Membrane-
Processes

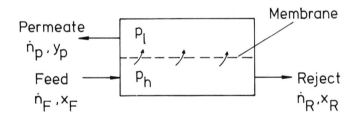

Figure 7 : Schematic Description of a Membrane - Module

It means that the local separation efficiency of the process varies
continuously along the membrane. For the mathematical modelling of
the behaviour of such an apparatus, it means that the permeate
flow n_p leaving from the module and its composition can only be
calculated in that the mass quantities dn_i permeating as per Eq.
(3) at a definite point through an infinitesimal small surface
element dA per time unit are summed up by integration via the
total membrane area.

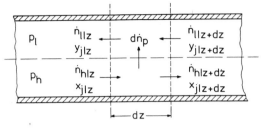

Figure 8 :
Balance of Mass Flow in a
differential Volume- Element

Hence it gives

$$dn_i = J_i * dA = (Cp_i/S)*(p_h * x_{i(z,b)} - p_l * y_{i(z,b)})*dA \qquad (8)$$

$$\text{with} \qquad dA = dz * db \qquad\qquad\qquad (9)$$

and the length z and the width coordinate b of the membrane area as
well as $x_{i(z,b)}$ and $y_{i(z,b)}$ as molar fractions of component i on
the high- and low-pressure side of the membrane.
Balance equations at the infinitesimal small volume element of
place z, Fig. 8, as well as macroscopic balances on the entire
separation unit, Fig.7, give more or less complex differential
equations wich usually can be solved numerically /11,13/.
In setting up the balance equations, a lot of important inner and
outer parameters and factors of influence are to be taken into
account:
- Construction type of module : capillary, plate-and-frame,
spiral-wound module, see Fig. 5
- Type of flow : cocurrent, cross, countercurrent flow of gases
on the high- ·and low-pressure side of the membrane, see Fig.5
with laminar or turbulent velocity distribution. Above all in
capillary and spiral-wound modules it is necessary to consider the
often appreciable pressure loss of the flow on one or both sides
of the membranes which will anyhow affect the efficiency of the
process (enrichment, depletion, recovery) /4,8/.
- Assumptions on the nature of concentration distribution at
either side of membrane (ideal mixture across the entire membrane
surface, no concentration polarization etc.). Besides, for certain
operation states, effects resulting from back-mixing shall also be
considered (back-diffusion, convections occurring by density or

temperature gradients, etc.) /13/.

- Composition of gas mixtures to be separated (two or more components)/13a/.
- Single or multi-stage circuits of modules, Fig. 6, with or without feedback of partial flows.

Keeping a single module and a binary gas mixture as an example, then the following dimensionless parameters may be defined :

- pressure ratio $\qquad P_r = p_l/p_h$ 　　　　　　　　(10)
- mass flow ratio
  - high pressure side $\qquad q = n_h/n_F$ 　　　　　　(11)
  - low pressure side $\qquad \theta = n_l/n_F$ (stage cut) (12)
- dimensionless membrane area
  $\qquad A' = Cp_A*p_h*A/n_F*S$ 　　(13)
- selectivity $\qquad \alpha = Cp_A/Cp_B$ 　　　　　(14)

Then Eq. 8 can be transformed to a term in the form of

$$dq/dA' = d\theta/dA' = f(x, y, Pr, \alpha)$$ 　　　　　(15)

For the concentration variations on both high- and low-pressure sides, the following terms will generally result :

$$dx/dA' = f(x, y, Pr, q, d^2x/dA'^2, dq/dA')$$ 　　(16)

$$dy/dA' = f(x, y, Pr, \theta, d^2y/dA'^2, d\theta/dA')$$ 　　(17)

The coupled system of differential equations can be solved numerically. The solutions naturally vary depending on geometrical or other requirements incorporated in the basic equations. It gives the specific solutions for the following correlations :

$$x_R = f(Pr, \alpha , \theta)$$ 　　　　　　　(18)
$$y_P = f(Pr, \alpha , \theta)$$ 　　　　　　　(19)
$$A' = f(Pr, \alpha , \theta)$$ 　　　　　　　(20)

with parameters of Eq. 10 - 14. To design a module or a plant, one will employ the solutions shown in graphical form on Fig. 9 and 10.

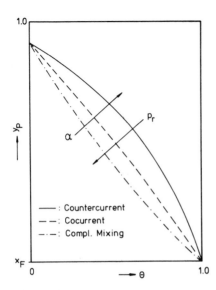

Figure 9 :

Results of theoretical calculation: Effect of α , Pr and various designs on the mole-fraction
y_P of the permeate stream /11/

Figure 10 :

Results of theoretical calculation: Effect of α , Pr on mem-
brane-area for the countercurrent case /11/

With preset values of x_F and Pr, given α and wanted permeate concentration y_P it follows a specific cut θ , Fig. 9, depending on
the type of flow. From Fig. 10 follow the concentration of reject
x_R and the dimensionless value of A', which allows for calculating
the membrane area A needed for this separation process.
Theoretical calculations and experimental results coincide largely
if in the basic differential equations all denominated conditions
are duly considered. This is reflected by Fig. 11 and 12; experiments and calculations were carried out by Kothe /13/ and Wonsak
/14/ at our Institute.
Besides the above mentioned quantities of design-data, the variables of recovery R and separation factor T can be specified which
are of prime importance for comparative process studies :
Recovery :

Permeate :		$R_P = y_P {*} n_P / x_F {*} n_F$	(21)
Reject :		$R_R = x_R {*} n_R / x_F {*} n_F$	(22)
Separation-Factor :		$T = (y_P/(1 - y_P)){*}((1 - x_R)/x_R)$	(23)

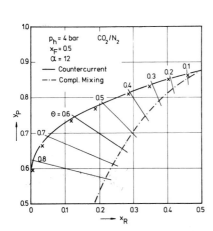

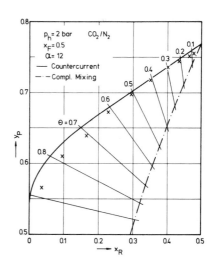

Figure 11 :
Comparison of theor. calculation
and experimental results;
capillary module /13/

Figure 12 :
Comparison of theor. calculation
and experimental results;
plate- and frame- module /14/

Thereby, you can manage to describe and optimize not only the construction of single modules but also that of multi-stage membrane
plants, that means both in regard to the choice of circuit type of
single commercial modules each with a known stage-characteristic
and with regard to the utilization of certain process concepts,
such as offered by the "Continuous Column"- Technique as per Hwang
/15/ or with modifications as per Schulz /16/.

As the gas separation process by means of membranes belongs to the
group of pressure driven processes, the level of the feed pressure
P_h and the pressure ratio Pr bear prime importance in assessing
the process profitability, because, apart from capital cost for
compressors and modules, the energy cost of the process must also
be taken into account. These are proportional to the compressor
power

$$N \sim n_F * p_h \qquad\qquad (24)$$

and rise even more if a process must be run in multi-stage operation with intermediate compression. Definitive data on this capital

and energy cost can only be derived from concrete examples. It will hardly be possible to determine common data bearing in mind the variety of parameters, Eq. 10 - 14, construction types and circuit types.

But some criteria can indeed be discussed :

-   If the selectivity of a membrane is high, the product purity on the reject- and permeate-side, respectively, can be raised, and the recovery alike, Eq. 21, 22. But an analysis /17/ shows that in case that α becomes higher than 20, the pressure ratio Pr will play a much more important part for the variables under discussion, i.e. x_R, y_P, R_P, R_R, not least for cost considerations.

-   With a given selectivity and n_F, Eq. 13 according to which the necessary membrane surface A can be calculated (depending on various parameters) shows, at least by tendency, that A can be reduced provided that

. the Cp-value could be increased,

. the active layer thickness S could be reduced,

. the pressure on the feed side would be raised.

The first two points may be fulfilled by searching out a better polymer or an ingenious manufacturing of the membrane. The latter will rapidly hit against economic limits, because, as per Eq. 24, the compressor power to be installed will rise. Therefore it will be the tendency to keep the feed gas pressure as low as possible /6/. The membrane technology will work particularly favourable in terms of profitability, also compared with alternative unit-operations, if the modules can be charged with process gas which is already set at high pressure (H_2 from NH_3 synthesis, CH_4 from natural gas fields), and if the produced gas can be exploited and transported without additional compressors to be installed /3,8,17a/.

-   Pressure losses occurring due to the gasflow along the membrane with or without spacers at the high- and the low-pressure side should be minimized. This actually promises to bear some advantages for the plate-and-frame concept and, respectively, it leads to the idea to build the capillary modules with not too narrow capillary diameters and/or short lengths. (For optimal diameter/length ratio see /11/.) The latter idea has obviously been persued by Dow /6/.

Economy of Membrane Processes - Comparison with Alternative
Technologies

The above mentioned membrane process is not the only chemical-engi-
neering technology suitable to separate gas mixtures; for numerous
years, there have been known some other unit-operations which
already reached a highly sophisticated technical level, i.e. the
operations of

- cryogenic
- adsorption
- absorption.

Before a commercial user will decide to purchase a feasible technology, he will check the alternatives for various criteria. Frequently, it is the economy which is of prime importance.
In hydrogen recovery from chemical process gases, the high level
of economy of the very simple and easy to operate membrane facilities has already been proved and demonstrated /3,4/. For other
separation problems, the large-scale technical employment of membranes is about to reach a breakthrough, in some other cases it
will still take several years of intensive development work until
the rather conservatively biased big industry will accept the "New
Technology". The acceptance, as said before, can only be reached
and attained by rendering proof of higher efficiency and profitability of membrane technology.
To put some examples, we will have a look at the CO_2/CH_4 - separation and the oxygen-enrichment from air.
As for the former problem, a great number of profitable studies
have been made in recent years proceeding from different viewpoints
/see 18/. We ourselves in cooperation with an engineering company
(C. Still, Recklinghausen, FRG) have endeavoured to design a
medium-scale plant for the purification of landfill gas /18/. Our
results, in regard to the costs of separation of biogases, corre-
spond with the calculation made by Finken /19/, see points A and F
of Fig. 13. The height of the histogram beams of this figure indicate the costs of gas purification at different volume flows of
crude gas for appropriate membrane plants and absorptive pres-
sure-water- and amine-scrubbing.
Accordingly, the membrane process is in principle indeed competitive, at least within the range of discussed volume flows. Even in
those cases where H_2O, H_2S, polychlorinated or higher hydrocarbons

58

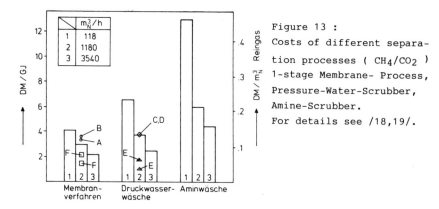

Figure 13 :
Costs of different separa-
tion processes (CH_4/CO_2)
1-stage Membrane- Process,
Pressure-Water-Scrubber,
Amine-Scrubber.
For details see /18,19/.

occur in the crude gas which possibly have to be separated prior to
entering into the actual gas separation process by means of other
techniques, the costs will shift only to a negligible extent and
also hardly to the disadvantage of the membrane process, because
the other processes need to take similar arrangements or, as in
case of CA- membrane /20/ H_2O and H_2S, may be separated very
favourable and with high efficiency from the desired product CH_4.
For more details see /18,19/.

For a lot of chemical processes and for those employed in thermal
energy recovery, combustion air enriched to approx. 30% O_2 would
appreciably raise the efficiency of these plants. Even today, potential or already proposed membrane processes compete directly
with the cryogenic technology (for larger quantities) and with
the wide-spread pressure-swing adsorption on zeolites or similar
molecular sieves. Tippmer /21/ has calculated for these cases the
cost per m^3 oxygen as a function of the product quantity per hour.
Fig. 14 summarizes the results.

The required combustion air containing 30% O_2 is generated by
mixing ambient air with O_2 from pressure bottles ($O_2 > 99,5$% vol)
for consumptions $<$ 800 m^3/month, and from a cold-vaporizer for
consumption rates $>$ 800 m^3/month. For quantities above approx. 700
m^3/h, a cryogenic plant with price-efficient optimal 80% O_2 feeds
the mixing unit. The mentioned pressure-swing adsorption works with
zeolites. Membrane facilities were calculated in different arrangements :

- Mem B with Cp/S = 16,4 Barrer/μm, α = 4,8; membrane-rectifica-
  tion, see Fig. 6.

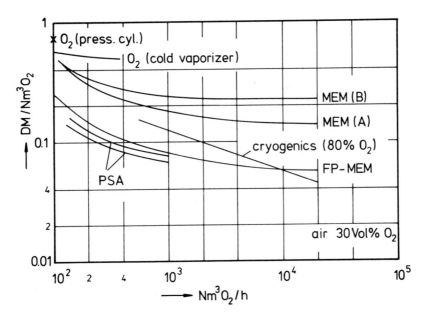

Figure 14 :
Cost of preparing combustion-air with 30% O_2 . Compared are preparation technics from High-pressure-cylinders, Cold-vaporizer, Cryogenics, Pressure-swing-adsorption and different Membrane-installa-
tions /21/

- Mem A with Cp/S = 130 Barrer/ μm, α = 2,2; single-stage cascade, see Fig. 6.
- FP - Mem as fictitious polymer membrane, single-stage   cascade
  with   Cp/S = 740 Barrer/ μm, α = 60

(1 Barrer = 10^{10} $cm^3 * cm/cm^2 * sec * cmHg$)

The membrane plants are designed such as to ensure that the permeate flow supplies the combustion air being enriched to 30% O_2.

As may be gathered from Fig. 14, no alternative technique is known
that competes with the cryogenic plant in the case of major quantity flows of O_2. For minor or medium quantities, conventional
membranes of type A or B can hardly compete with pressure-swing ad-

sorption processes. Only membranes of type FP with currently still
utopic properties could reach the cost level of alternative processes.

Corresponding calculations for an oxygen enrichment up to 90 or 95%
still indicate substantial distances between membrane and classical
technologies /21/.   Not least Japan's industry set its goal in a
MITI-program with the aim to quickly elaborate efficiently working
membrane systems for oxygen enrichment from ambient air /5/. It
will be quite interesting to observe the contest of industry giants
fighting for this huge market in the years ahead.

While oxygen can hardly be enriched more economically or efficiently
than by means of conventional processes within a foreseeable time,
the recovery of nearly pure nitrogen from ambient air ($>$ 98%), at
least in quantities $<$ 1,000 m^3/h for inertialization (blanketing)
purposes may indeed prove to be an economically efficient and
profitable process /6/.

Prospects

The theoretical calculations and design of membrane modules and
entire plants based on known outer conditions (pressure, temperature, mass flows), membrane properties as well as given and demanded concentrations of feed, reject and permeate have already
become an ascertained level and are state-of-the-art.

The crux for a great deal of commercially interesting applications
of this technology lies in the yet unsatisfactory properties of
known membranes :

                lack of selectivity at high permeability.

Obviously, however, as indicated by publications in the past two
years, the US and Japanese industry are working at high pressure on
innovations which in classical manner favourize homogeneous /6/ or
composite-membranes /5,7/ with a very thin active separation layer.

Besides, it should not be ignored that there are some "alternative"
concepts on membrane and, respectively, membrane-systems which
tender themselves for the "Next Generation" :

- Extremely thin, "pin-hole"-free skin layer are plasma-polymerized from optional monomer.gas phase on a highly-porous support layer /22/.

- The thinnest of all conceivable membranes are monomolecular thick. Straightened and cross-linked "monolayers" bound to a porous support, possibly of biological origin, appears to be producable /23/. As one monolayer alone, without specific "pores" or "geometric structures" does not necessarily represent a selective separation membrane, the realization of this principle still is a problem of basic research and technological development which takes more time.

- Gases diffuse in liquids by 2 to 4 powers faster than they do in polymers. Nothing better than to utilize these properties and to fill the void of porous membranes with a highly selective liquid or immobilize the liquid in the gel-like amorphous structure of an appropriate polymer. Even the swelling of a homogeneous polymer with a softener like H_2O, glycerol etc raises the permeability appreciably, but not necessarily the selectivity /21/.
The drawbacks of these immobilized-liquid-membranes are presently still predominant as they are generally relatively thick (20 to 200 μm or more), whereby the merit of the high permeability versus the mere polymer membranes with a thin skinlayer is lost again. The volatility of liquids and the restricted pressure strength (capillarity) limit the service life; also the module construction itself is extremely difficult /24/.

- Nevertheless, this last mentioned technology appears to be quite interesting for the future because "carriers" can be added to the liquid phase which, above all, are capable of raising the selectivity of this system enormously /25/.

In view of the latter discussed aspects, too, seen from the engineering side of the coin, the gas separation by means of membrane continue to be a problem of developing appropriate membrane systems rather than a problem of the technical realization of modules or plant units, no matter how difficult the latter might be in a single case.
Consequently, the polymer- and physical chemists of all nations are called upon to throw themselves into a contest under the world-wide

known Olympic slogan :

ALTIUS - CITIUS - FORTIUS

They may create membranes which are

- HIGHER in selectivity
- FASTER in permeability
- LONGER in the jumps

towards a successful future of "High-Tech-Gas-Separation" by means
of membranes.

Literature:

/1/ Graham, Th.: Philos. Magazine and Journal of Science, Series 4,
32(218), 401 (1866)

/2/ Loeb, S., S.Sourirajan : Report Nr.6060,Dep. of Engineering,
University of Cal., LA, CA, 1960

/3/ Monsanto Company, div. techn. informations, Monsanto Europe,
Brussels, Belgium

/4/ Kurz,J.E.; R.S.Narayan : New Developments and Applications in
Membrane Technologie; paper pres. Apr. 1986, AIChe Spring Meeting,
New Orleans, Lou.

/5/ Nakamura, A.,M.Hotta : Chem. Econ.& Engin. Review 17(1985)N.7/8

/6/ Bantinidis,D., H.Dittrich : Generon Air Separation System, Dow
Chemical Germany

/7/ Europe Patent 0 099 432 v.1.2.1984, Toray-Industries, Japan

/7a/ Eickmann, U. et al. : German Chem. Eng. 8(1985) 186/194

/8/ Rautenbach, R.: Chem.Ing.Techn. 57(1985)2; 119/130

/9/ Hwang,S.T., C.K.Choi, K.Kammermeyer : Separation Science 9(6)
(1974) 461/478

/10/ Strathmann, H. et al.: Chem.Ing.Tech. 57(1985)7, 581/596

/11/ Hwang, S.T., K. Kammermeyer : Membranes in Separation, Techniques of Chemistry, Vol.VII, J.Wiley + Sons, 1976

/12/ Rautenbach R, R.Albrecht : Membrantrennverfahren, Verlag Sarre
+ Sauerländer 1981

/13/ Kothe, K.D., H. Wonsak : Einfluß der Rückvermischung auf das
Trennverhalten von Gaspermeatoren. Paper pres. at Dechema Working-
party "Membrantechnik", Jan.1986, Frankfurt FRG

/13a/ Rautenbach, R., W. Dahm : Chem.Eng.Process. 19(1985)211/219
/14/ Wonsak, H. : Unpublished Data From a Plate-and-Frame Module, University of Dortmund, Mechan.Verfahrenstechnik, FRG
/15/ Hwang, S.T., J.M.Thorman : AIChe J. 26(1980) 558 ff
/16/ Schulz, G. : Investigations on Gas-Separation by means of membranes, Dr.-Thesis, University of Dortmund 1983
/17/ MacLean, D.L., D.J. Stokey, T.R.Metzger : Hydrocarbon Proc. 62(1983) 47 ff
/17a/ McCandless,F.P. :J.Membr. Sc. 24(1985) 15/28
/18/ Werner, U. : GWF - Gas/Erdgas , 126 (1985)1, 25/29
/19/ Finken, H., L.Belau : Chem.Ing.Tech. 56(1984)12, 944/945
/20/ Grace Membrane Systems, Technical Informations, W.R. Grace + Co, Columbia, MD
/21/ Tippmer, K. : Modification of Membranes for Gas-Separation, Paper pres. Jan.1986, Dechema, Frankfurt, FRG; Working Party "Membrantechnik"
/22/ Osada, Y. : Membrane 10(1985) 215/223
/23/ Sara, M., U.B. Sleytr : Isoporous ultrafiltration membranes from crystalline bacterial envelope layers. Paper presented at the 5. Symposium on Synthetic Membranes in Science and Industrie, Sept. 1986, Tübingen, FRG
/24/ Friesen, D.T., W.C. Babcock : Novel applications of supported-liquid membranes, Paper pres. in Tübingen, see /23/
/25/ Noble, R.D., J.D.Way : Facilitated Transport of acid gases in Ion-exchange membranes, Paper pres. in Tübingen, see /23/

A Unified Theory of Separation Processes Based on Irreversible Thermodynamics

By R. Krishna

INDIAN INSTITUTE OF PETROLEUM, DEHRA DUN 248 005, INDIA

Abstract

The theory of Irreversible Thermodynamics (IT), in particular the Generalized Maxwell-Stefan relations for describing the relative motion of species in a multicomponent mixture, is used to develop the fundamental basis for separation processes. The IT approach, and formulation, is shown to be indispensible in the estimation of the mass diffusivities of thermodynamically non-ideal fluid mixtures, description of separation processes in the region of the critical point, prediction of mass transfer rates and stage efficiencies in multicomponent separation processes, and for the correct modelling of separation processes involving more than one driving force. Use of the "dusty" fluid model allows the IT formulation to be consistently extended to the treatment of separation processes involving porous barriers or membranes.

1. Introduction

There are two basic steps involved in developing a separation process:

(1) The first step is to effect _relative_ motion of the molecular species present in the mixture to be separated. To achieve this, a force must be exerted on each of the species; the various separation possibilities differ in the type of force exerted and by the manner in which this force is "created".

(2) The next step is to determine the rates at which the relative motion takes place. The knowledge of these rates is essential in the sizing of the separation equipment.

The theory of Irreversible Thermodynamics (IT) identifies the various driving forces which can cause relative motion of molecular species. Also, the IT formulation provides a relationship between the driving force exerted on the species and the velocities at which these species are caused to move in relation to the mixture. IT therefore offers the potential of being used to develop a unified theory of separation processes.

The starting point in the theory of IT is the expression for the rate of entropy production caused by the relative motion of species set up during separation (see [1,2] for derivation):

$$\sigma = -\frac{1}{T} \sum_{i=1}^{n} c_t RT \, \underline{d}_i \cdot (\underline{u}_i - \underline{u}) \geq 0 \tag{1}$$

where $c_t RT \, \underline{d}_i$ represents the driving force exerted on species i _per unit volume of the mixture_, which tends to move species i _with respect to the mixture_ with a relative velocity given by $(\underline{u}_i - \underline{u})$. This relative motion is the essence of separation. In the absence of any driving force there is no relative motion between the various species; the mixture is at thermodynamic equilibrium:

$$\sigma = 0; \quad \underline{d}_i = 0 \qquad \text{(thermodynamic equilibrium)} \tag{2}$$

When finite driving forces $\underline{d}_i$ are exerted on the species, the rate of entropy production is positive definite (cf. Eq. (1)), signifying the fact that energy must be expended in achieving separation. It remains to identify the various driving forces $\underline{d}_i$ and to relate the $\underline{d}_i$ to the $(\underline{u}_i - \underline{u})$.

64

2. Driving Forces and Constitutive Relations

From the theory of IT the driving force for relative motion of molecular species i with respect to the mixture is seen to be the sum of three constituent driving forces:

1) driving force arising out of chemical potential gradients which are created in the system; these forces are <u>always</u> present in separation processes even though they need not be the "initial" cause for separation to occur.

2) driving forces arising out of pressure gradients generated in the system. For pressure gradients to be effective in separation of a mixture there must be a difference in molecular weights of the species to be separated or put another way, there must be difference between the volume fraction and the mass fraction of the mixture.

3) for electrically charged species, application of an electrostatic potential gradient will cause the species to be set in relative motion.

The general expression for the overall driving force $\underset{\sim}{d}_i$ has been conveniently put in the following form by Lightfoot [1]:

$$\underset{\sim}{d}_i = \frac{x_i}{RT} \underset{\sim}{\nabla}_{T,p}\mu_i + \frac{(\phi_i - \omega_i)}{c_t RT} \underset{\sim}{\nabla}p + x_i z_i \frac{F}{RT} \underset{\sim}{\nabla}\phi \tag{3}$$

In a mixture of uniform composition there will be no chemical potential gradients. To this mixture the application of a pressure gradient, for example, will set up a finite driving force $\underset{\sim}{d}_i$ (provided the volume fraction ϕ_i differs from the mass fraction ω_i) causing relative motion of molecular species. This motion in turn will cause chemical potential gradients to be created which will act in a way to counter the pressure gradient till a dynamic equilibrium is reached (with $\underset{\sim}{d}_i$ = 0). There is a useful composition difference which exists at this equilibrium condition and separation is achieved.

Having identified the three basic constituents of the driving force $\underset{\sim}{d}_i$, it now remains to relate the (relative) velocities of motion of the species to this driving force. This relationship, called the constitutive relationship, is most conveniently written in the following two <u>equivalent</u> forms (see Lightfoot [1] and Standart, Taylor and Krishna [2] for derivation):

$$\underset{\sim}{d}_i = \sum_{\substack{j=1 \\ j\neq i}}^{n} \frac{x_i x_j (\underset{\sim}{u}_j - \underset{\sim}{u}_i)}{\mathcal{D}_{ij}} \equiv \sum_{\substack{j=1 \\ j\neq i}}^{n} \frac{(x_i \underset{\sim}{N}_j - x_j \underset{\sim}{N}_i)}{c_t \mathcal{D}_{ij}} , \quad i = 1,2,..n \tag{4}$$

The constitutive relation (4) is called the Generalized Maxwell-Stefan (GMS) equation and is, in my opinion, the most convenient flux-driving force relation from a practical point of view. Equations (3) and (4) form the basis of our unified approach to the theory of separation processes. In their complete form they do appear awesome and before proceeding further it is instructive to deal with a few limiting cases, in order to emphasise the utility of the generalized approach.

First, let us consider the simplest case of separation of a binary mixture in the absence of pressure gradients and electrostatic potential gradients. The GMS equations take the form:

$$\underset{\sim}{d}_1 = \frac{x_1}{RT} \underset{\sim}{\nabla}_{T,p}\mu_1 = (1 + x_1 \frac{\partial \ln\gamma_1}{\partial x_1}) \underset{\sim}{\nabla}x_1 = -\frac{\underset{\sim}{J}_1 \equiv \underset{\sim}{N}_1 - x_1(\underset{\sim}{N}_1 + \underset{\sim}{N}_2)}{c_t \mathcal{D}_{12}} \tag{5}$$

or

$$\underset{\sim}{J}_1 = -c_t \mathcal{D}_{12} \Gamma \underset{\sim}{\nabla}x_1 = -c_t D_{12} \underset{\sim}{\nabla}x_1 \tag{6}$$

where D_{12} is the Fickian diffusion coefficient which is related to the GMS diffusivity $\mathcal{D}_{12}$ by

$$D_{12} = \mathcal{D}_{12} \Gamma; \quad \Gamma = (1 + x_1 \partial \ln\gamma_1/\partial x_1) \tag{7}$$

FICKIAN DIFFUSION COEFFICIENT

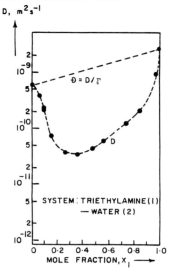

Fig. 1. Variation of the Fickian diffusivity D and the GMS diffusivity with composition. Dudley and Tyrrell [3].

The thermodynamic factor Γ is a strong function of composition and, as a consequence the Fickian diffusivity D shows a strong composition dependence. Figure 1 shows the strong variation of the Fickian diffusivity of triethylamine(1)-water(2). The GMS diffusivity $Đ_{12}$, calculated as D_{12}/Γ, shows a much simpler variation with composition. This goes to show that the GMS diffusivity more closely reflects the "kinetic" nature of the diffusion process whereas the commonly used Fickian diffusivity relects both the "kinetic" and "thermodynamic" factors. Separation of these two factors is useful from the point of view of calculation of the transfer rates. From Fig. 1 it can be seen that log $Đ_{12}$ varies with x_1 in a linear fashion. This (empirical) observation has been confirmed for a variety of systems by Vignes [4]. Acceptance of this allows calculation of $Đ_{12}$ from a knowledge of the infinite dilution diffusivities. The Fick's law diffusivity can then be calculated from the GMS diffusivity by use of Eq. (7). The use of the GMS constitutive relation, in preference to the conventionally used Fickian formulation, provides a convenient practical treatment of mass transport phenomena in non-ideal fluid mixtures.

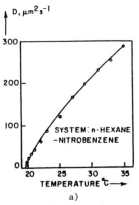

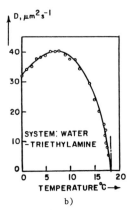

a)

b)

Fig. 2. (a) Fickian diffusivity of the system n-hexane – nitrobenzene as a function of temperature, x_1 = 0.58. (b) Fickian diffusivity of the system triethylamine – water, x_1 = 0.087. Data from Haase and Siry [5].

Another interesting observation concerns the variation of the diffusivity of a binary mixture with temperature in the region of the critical solution temperature (LCST or UCST), as shown in Fig. 2. The Fickian diffusivity falls sharply in value to zero as the CST is approached ! Again, with the application of IT theory this "strange" behaviour of the Fickian diffusivity can be understood. The reasoning is as follows. From considerations of thermodynamic stability, the determinant of the Hessian matrix $|G|$ must be positive definite (see Modell and Reid [5a]for more detailed discussions), i.e.

$$|G| \geq 0 \tag{8}$$

where the elements of the matrix $[G]$ are given by:

$$G_{ij} = \frac{\partial(\mu_i - \mu_n)}{\partial x_j} = G_{ji} = \frac{\partial(\mu_j - \mu_n)}{\partial x_i} = \frac{\partial^2 G}{\partial x_i \partial x_j} \tag{9}$$

where G is the molar Gibbs free energy. At the critcal point itself we have

$$|G| = 0 \tag{10}$$

It follows from Eq. (7) that the Fickian diffusivity must vanish at the critical point for a binary system. A few separation processes, such as liquid-liquid extraction could operate close to the critical point (the plait point in L-L systems) and the IT formulation is indispensible for the estimation of the transfer rates. The recent paper by Krishna et al [6] emphasised the need for using a rigorous IT formulation in describing the interphase mass transfer rates in L-L extraction.

Having demonstrated the utility of the GMS formulation in describing mass transport in non-ideal fluid mixtures, and in the region of the critical point, let us turn our attention to ideal gas mixtures. In this case the GMS Eqs (3) and (4) reduce to

$$\underset{\sim}{d}_i = \nabla x_i + \frac{(x_i - \omega_i)}{p} \nabla p = \sum_{\substack{j=1 \\ j \neq i}}^{n} \frac{x_i N_j - x_j N_i}{c_t \, Ð_{ij}} \tag{11}$$

where we have omitted the electrostatic potential gradient term, usually not of importance in separation. Also, for ideal gas mixtures the mixture molar density $c_t = p/RT$ and the volume fraction ϕ_i equals the mole fraction x_i. For ideal gas mixtures the GMS diffusivity $Ð_{ij}$ equals the Fickian diffusivity of the corresponding binary pair and these are composition independent. Equations (11), which are called the Maxwell-Stefan equations, are consistent with the kinetic theory of gases. The Maxwell-Stefan equations (11) are thus a special case of the IT theory.

It is important to appreciate that even for mixtures of ideal gases the diffusivities of the binary pairs $Ð_{ij}$ are, in general, unequal to one another. In fact in the sweep diffusion process for separation, as we shall see later, it is the difference in the diffusivities of the binary pairs which is harnessed for the purposes of achieving separation. The IT formulation takes this effect into account quite routinely. In the modelling of the mass diffusion process it is quite tempting to use Occam's Razor "Pluralitas non est ponenda sine necessitae" or in free translation "Do not complicate things beyond necessity". With this approach let us assume all pair diffusivities are (nearly) equal to one another leading to the simplification (cf. Eq. (11)):

$$\underset{\sim}{J}_i \equiv \underset{\sim}{N}_i - x_i \underset{\sim}{N}_t = - c_t \, Ð \, \underset{\sim}{d}_i, \quad i = 1,2,\ldots n \tag{12}$$

or in other words the diffusion flux J_i of component i, with respect to the mixture is proportional to its own driving force $\underset{\sim}{d}_i$. Duncan and Toor [7] studied diffusion in the ideal gas system $N_2 - H_2 - CO_2$ in a two-bulb diffusion

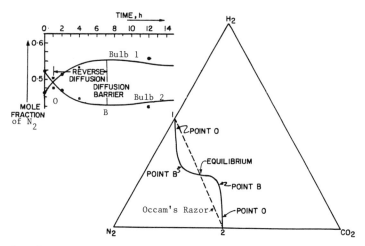

Fig. 3. Diffusion in a two-bulb diffusion cell (Bulbs 1 and 2). For the
particular starting compositions 1 and 2 denoted on the triangular diagram,
N_2 experiences osmotic diffusion at O, diffusion barrier at B and reverse
diffusion between O and B. The dashed lines ----- denote the composition
path that would have been followed if Occam's Razor Eq. (12) were to be
employed. Data of Duncan and Toor [7].

cell. Of particular interest are the composition profiles of N_2, shown in the
inset to Fig. 3. At point O, the compositions of N_2 in the two bulbs are the
same; nevertheless the diffusion of N_2 continues to take place. Between the
points O and B the composition of N_2 in bulb 1 continues to increase at the
expense of the composition in Bulb 2; in this region diffusion of N_2 takes place
from a region of low composition (bulb 2) to a region of higher composition
(bulb 1). At point B the composition profiles in both bulbs 1 and 2 are flat,
which signifies that no diffusion takes place at B despite the existence of a
large composition difference driving force. The three peculiar phenomena noted
have been respectively termed osmotic diffusion (at point O), reverse diffusion
(between O and B) and diffusion barrier (at point B). Beyond point B, diffusion
of N_2 takes place "normally". The diffusion behaviour of this _ideal_ gas mixture
cannot be described by use of the Occam's Razor model Eq. (12), wherein each
flux is taken to be proportional to its own driving force, with equal diffusion
coefficient for each component. As shown by Duncan and Toor [7], the Maxwell-
Stefan Eq. (11) is able to "model" the diffusion process quantitatively. In
separations involving gaseous mixtures with widely varying molecular weights the
complete form of the MS equations must therefore be used. Generally speaking,
gaseous mixtures with H_2 as one of the components will fall into this category.
    One further point regarding the two-bulb diffusion experiment of Duncan
and Toor [7] needs special mention. Does the phenomenon of reverse diffusion
violate the second law of thermodynamics ? The answer is an emphatic NO. The
positive definite condition for the rate of entropy production, Eq. (1), only
requires the total system entropy production to be positive definite. N_2 can
consume entropy in experiencing reverse diffusion _provided_ the other two species
H_2 and CO_2 produce entropy at such a rate that the total rate of production for
the system as a whole is positive. These two species are in effect pumping N_2
uphill. IT helps in rationalizing the experimental observations.

If in the two-bulb diffusion experiment, the transfer of N_2 were to be modelled using the "effective" diffusivity concept, then this effective diffusion coefficient for N_2 will assume values ranging from negative (in the region of reverse diffusion) to zero (at the point of diffusion barrier). Such "odd" behaviour of the effective diffusivity is characteristic of multicomponent systems, i.e. sytems having three or more species. The GMS pair diffusivity $Ð_{ij}$ is always positive definite. This can be seen by combining Eq. (1) and (4) which gives the following expression for the rate of entropy production:

$$\sigma = \frac{c_t R}{2} \sum_{i=1}^{n} \sum_{\substack{j=1 \\ j \neq i}}^{n} \frac{x_i x_j}{Ð_{ij}} (\underline{u}_i - \underline{u}_j)^2 \geq 0 \tag{13}$$

The positive definite condition for σ can only be met if the $Ð_{ij}$ are individually positive definite, i.e.

$$Ð_{ij} \geq 0 \tag{14}$$

The neat and compact form of Eq. (13), which contains no thermodynamic factors, also suggest the fundamental and superior basis of the GMS formulation of IT theory.

Having set up the theoretical framework for dealing with separation processes, let us first attempt to classify separation processes in a logical and systematic manner using this IT framework.

3. Classification of Separation Proceses

Relative motion between the species in a mixture can be caused by the action of one, or more, of the following (cf. Eq. (3))
   1) chemical potential gradients
   2) pressure gradients
   3) electrostatic potential gradients (effective only for charged species)
The above gradients may be created or made to act upon a system consisting of:
   - a single phase (G, L or S), or
   - a two phase system (V-L, G-L, L-L, G-S, L-S, G-PS, L-PS), or
   - a system consisting of two phases separated by a porous barrier or
     membrane (Fluid - M - Fluid; where Fluid = G, V or L)
where we use the notation: G = gas; V = vapour; L = liquid; S = solid; PS = porous solid; M = membrane or porous barrier.

The first scheme of classifying separation processes is based on the type of gradients (chemical potential, pressure or electrostatic potential) created in the system and whether these gradients are created in a (i) single phase, or (ii) a two phase system, or (iii) a system consisting of two fluid phases separated by a membrane or porous barrier.

The second scheme of classifying separation processes is according to whether the primary cause for separation is difference in the system (usually two-phase) composition at equilibrium or whether we rely on differences in transport rates of the species to achieve the desired separation. Thus we have:

(i) Equilibration Separation Processes. Here we have a useful composition difference between the two phases at equilibrium ($\sigma = 0$; $\underline{d}_i = 0$) and our efforts are devoted to promoting the approach to equilibrium. It is useful in predicting the actual approach to equilibrium (by use of Eqs. (3) - (4)), i.e. in predicting stage efficiencies or heights of transfer units. By operating industrial contactors with a high degree of turbulence, the rates of equilibration can be enhanced because of the additional, parallel, mechanism of turbulent mass transport.

(ii) Rate Governed Separation Processes. Here we have no separation occurring at equilibrium (e.g. a single phase system of homogenous composition at equilibrium or it may be that an unfavourable equilibrium exists in a heterogenous system. This situation is modified by the rate process so that we operate in a manner to prevent equilibration. The key to the separation is the

different rates of transfer of the component species in the non-equilibrium
situation. Essentially we rely here on the differences in the fluxes N_i created
within a single phase or across phase boundaries. IT plays a vital role here in
the calculation of the transfer fluxes N_i, by use of Eqs. (3) and (4). If we
were to rely only on the differences in the GMS pair diffusivities to ensure
differences in the rates of transfer N_i, then the separation factors α_{ij}:

$$\alpha_{ij} = (y_i/y_j)/(x_i/x_j) \tag{15}$$

(here y and x denote mole fractions in the two product streams) will be close
to unity (except in exceptional cases). Therefore in practice the separation
factors (or selectivities) of the process are enhanced by use of selective
barriers (or membranes) which allow predominantly the passage of one component.
The membranes may take different forms: porous diaphragm, metal perforated
screens, polymeric films with or without charged species, liquid surfactant
films etc.

Combining the two classification schemes above we may cite the following
examples of industrial importance:

Single Phase Separation Processes
- Equilibration Separation Process: centrifugation
- Rate Governed Separation Process: electrolysis, thermal diffusion

Two-Phase Separation Processes
- Equilibration Separation Process:
    V-L: distillation, partial condensation
    G-L: absorption, desorption, stripping, evaporation, gas extraction
    L-L: extraction
    G-S: sublimation, desublimation
    L-S: crystallization, zone melting, freezing
    G-PS: adsorption
    L-PS: adsorption, ion exchange, leaching
- Rate Governed Separation Process:
    condensation of azeotrope vapours through inert gas; evaporation of
    azeotrope liquid into an inert gas

Membrane Separation Processes
- Equilibration Separation Process:
    osmosis
- Rate Governed Separation Process:
    gas diffusion through barriers, mass or sweep diffusion, permeation of
    fluids through polymeric films or liquid surfactant films, reverse
    osmosis, ultrafiltration, dialysis, electrodialysis, pervaporation.

We now take up the discussions of separations under the above three main
categories and shall see how IT can be useful in the modelling and design of the
separation process.

4. Single Phase Separation Processes
4.1 Centrifugation

Consider a binary system made up of uncharged species and subjected to a
pressure gradient. Provided the volume fraction ϕ_i is different from the mass
fraction, the two species will experience a different force (cf. Eq. (3)) and
relative motion between the species will result. Due to the composition difference arising out of this relative motion a chemical potential gradient will be
set up which will act in a "direction" tending to equalize the composition differences. Eventually a thermodynamic equilibrium condition will be attained
wherein the two forces (due to chemical potential and pressure gradients) will
balance each other. At equilibrium, $d_1 = 0$, the composition distribution of
component 1 will be given by (cf. Eq. (3))

$$\Gamma \frac{dx_1}{dr} = \frac{(\omega_1 - \phi_1)}{c_t \, RT} \frac{dp}{dr} \tag{16}$$

which shows that "dense" molecules for which the mass fraction is greater than the volume fraction will tend to move preferentially down the pressure gradient. For an aqueous solution of component 1, for example, an indication of the magnitude of the pressure gradient necessary to cause separation can be gleaned from the fact that c_t RT = 138 MN/m^2 and therefore pressure gradients of the order of a few thousand atmospheres must be set up across the sytem in order to achieve measurable separations. In practice, large pressure gradients can be developed in a centrifuge for which

$$\frac{dp}{dr} = \rho \, \Omega^2 \, r \tag{17}$$

where Ω is the angular velocity ($\Omega = 2\pi$ f where f is the rotational speed in revolutions per second) and r is the radius of the centrifuge. Many thousand rotations per minute are required to induce the necessary gradients of pressure.

For an ideal gas mixture, the separation factor α_{12} (defined by Eq. (15)) can be obtained by integration of Eq. (16)-(17) and works out to

$$\alpha_{12} = \exp \left[(M_1 - M_2)\frac{\Omega^2 \, r^2}{2RT} \right] \tag{18}$$

Ultracentrifugation is the industrially used technique for the separation of the isotopes of uranium: $U^{235}F_6$ (M_1 = 349.15) and $U^{238}F_6$ (M_2 = 352.15). Even at a rotational speed of 40,000 rpm the separation factor, calculated from Eq. (18) taking r = 60 mm, works out to only 1.0396. For commercial operation to achieve the desired throughput and separation, a few million centrifuges are required.

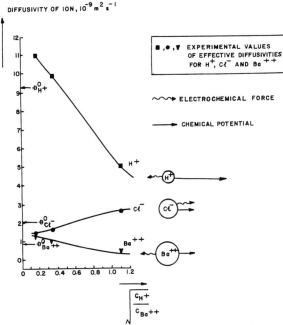

Fig. 4. Diffusion in mixed electrolyte systems. The effective diffusivity of an ion is strongly influenced by the electrostatic force "pull" or "push" exerted on it. Data from Vinograd and McBain [8].

4.2 Single Phase Separations involving Ionic Species

Consider an aqueous solution of electrolytes. The mixture consists of the ionized species (both + and - vely charged) and the "solvent" water. If a electrostatic potential is applied to the solution then the + ve ions will move towards the - ve electrode and the - ve ions will move in the opposite direction towards the + ve electrode. For <u>dilute</u> electrolyte solutions, Eqs. (3) and (4) reduce to the following expression for each ionic species i:

$$\underset{\sim}{N}_i = - c_t \, \theta_{in}^0 \, \underset{\sim}{\nabla} x_i - c_i z_i \, \theta_{in}^0 \, \frac{F}{RT} \, \underset{\sim}{\nabla} \phi + c_i \, \underset{\sim}{u}_n \qquad (19)$$

where the subscript n denotes the "solvent" water. Eq. (19) is usually referred to as the Nernst-Planck equation and as we see it is only a special case of the GMS formulation of IT. The charge on each ionic species, z_i, can be positive or negative (e.g. $z_{H}^+ = +1$; $z_{Cl}^- = -1$) and so the electrostatic force can act either in the same or opposite direction to the composition gradient.

Even when no electrostatic potential is imposed on the system, Eq. (19) must be used to describe ionic mass transport because when diffusion of ionic species takes place there will be a charge separation due to the fact that the ions have different intrinsic diffusivities (e.g. $\theta_{H}^+ = 9300$; $D_{Cl}^- = 2000$; $D_{Ba}^{++} = 850 \ \mu m^2 \ s^{-1}$). In the bulk solution there will be a tendency to maintain electroneutrality:

$$\sum_{i=1}^{n} c_i z_i = 0 \qquad (20)$$

To maintain electroneutrality an electrostatic force will be created on each of the ionic species. This effect can best be explained in terms of the data of Fig. 4. Here we see that the "effective" diffusion coefficient of H^+ ion is lowered considerably due to the presence of other ions. The electrostatic force tends to act as a "leash" tending to impede the transfer of H^+ in the interests of maintaining electroneutrality. On the other hand the relatively sluggish Cl^- is made to move "faster" by the electrostatic "pull". The, even more sluggish, Ba^{++} is slowed down even further by the electrostatic leash, as pictured in Fig. 4. For separation processes involving ionic species, such as ion exchange and metals extraction, the correct modelling of the transfer process using the IT formulation is essential. We shall be touching on this later on in this paper but the principles behind the phenomena are those described above.

4.3 Thermal Diffusion

In setting up the GMS Eq. (4) we had ignored the effect of thermal diffusion, i.e. diffusion induced by a temperature gradient. Thermal diffusion is usually unimportant in practice, but in a few cases this phenomenon can be used to effect separation. To take account of thermal diffusion, the term on the right hand side of Eq. (4) has to be augmented to include the effect of a temperature gradient. For a binary system, for example, we may write (cf. Eq. (6)):

$$\underset{\sim}{J}_1 = - c_t \, \theta \, \Gamma \, (\underset{\sim}{\nabla} x_1 + k_{1T} \, \underset{\sim}{\nabla} \ln T) \qquad (21)$$

where k_{1T} is the thermal diffusion ratio. The greater the value of k_{1T}, the greater is the separation achievable by thermal diffusion. When k_{1T} is + ve, species 1 moves down the temperature gradient; when k_{1T} is - ve, component 1 moves up the temperature gradient to the warmer region. k_{1T} is usually one or two orders of magnitude smaller than unity and therefore the separation factors achieved by thermal diffusion are close to unity. For successful practical application the small separations achieved can be enhanced by thermal convection as is done in the thermogravitational thermal diffusion column of Clusius and Dickel [9]. In gaseous mixtures at normal temperatures the heavier molecules usually diffuse down the temperature gradient, leading to a higher composition in the colder region.

The thermal diffusion ratio is more sensitive than any of the other transport coefficients to the <u>nature</u> of the intermolecular forces. Whereas viscosity, thermal conductivity and molecular diffusivity are first order effects depending primarily on the occurrence of molecular collisions and only secondarily on the <u>nature</u> of these collisions, thermal diffusion effect arises from a second order process and the values of the thermal diffusion coefficient may be positive, zero or negative according to the nature of the molecular interactions. Commercial application of the thermal diffusion phenomena is mainly for the separation of isotopes which show differences only at a second order level; see Rutherford [10].

5. Two Phase Separation Processes

In this class of separation processes two phases are created from the original mixture either by the input of energy (heating or cooling) or by the addition of a separating agent; this agent may take the form of a liquid, a gas or a porous solid. Most of the processes in this category are equilibration separation processes and thus rely on the fact that at equilibrium there is a useful composition difference between the two phases. For the two phases I and II, the condition of thermodynamic equilibrium, Eq. (2), leads to the condition:

$$\ln\left(\frac{\gamma_{iI}\,x_{iI}}{\gamma_{iII}x_{iII}}\right) + \frac{(\overline{V}_i)_{avg}}{RT}(p_I - p_{II}) + \frac{z_iF}{RT}(\Phi_I - \Phi_{II}) = 0 \qquad (22)$$

wherein we have noted that the volume fraction $\phi_i = c_i\overline{V}_i$ where $\overline{V}_i$ is the partial molar volume of species i. Equation (22) is the generalized expression for equilibrium between two phases, which follows nicely from a general IT theory. The treatment of phase equilibria and their prediction from a knowledge of the molecular properties and functional groups as discussed in the excellent texts of Praustnitz [11] and Reid, Prausnitz and Sherwood [12].

Most industrial contactors operate under steady-state conditions; here the gradients in composition in either phase are dictated by the operating conditions <u>and</u> the equilibrium composition distribution (satisfying Eq. (22)). The GMS formulation Eqs (3) – (4) is useful in determining the steady-state transfer rates and hence stage efficiencies.

In the class of two-phase separation processes distillation is by far the most widely used. It is surprising therefore that it is only recently that it has been realized that for highly non-ideal mixtures it is absolutely necessary to use the GMS formulation for estimation of the transfer efficiencies on trays. Figure 5 compares the experimentally measured composition profiles for the system acetone-methanol-water with the theoretical predictions based on two different approaches: (1) based on the GMS formulation of IT theory, Eqs. (3) – (4), and (2) based on the Occam's Razor "model" Eq (12), which is equivalent to assuming equal component transfer efficiencies for each individual component in the multicomponent mixture. The results given in Fig. 5 demonstrate the clear superiority of the GMS formulation of IT; see Krishnamurthy and Taylor [13].

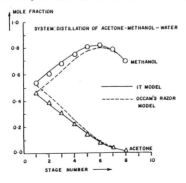

Fig. 5. Composition profiles in a distillation tray column. Comparison of the experimentally measured profiles with predictions of:
(1) ——— IT theory, Eqs (3)-(4)
(2) ------ Occam's Razor, Eq. (12)
Comparative study carried out by Krishnamurthy and Taylor [13]

The results of Fig. 5 only show the superiority of the IT formulation in
<u>simulation</u> of experimental distillation results. The effect of differing const-
ituent binary pair $Đ_{ij}$, in both vapour and liquid phases, can also be very significant in column design calculations. Krishnamurthy and Taylor [14] studied
four cases of column design using both the GMS formulation and the Equal Efficiency (i.e. Occam's Razor) approach. Their key results in respect of the number
of trays required for achieving a specified separation are tabulated below.

Table 1. Number of Distillation Trays Required for Specified Separation

System:	Equal Efficiency Assumption	GMS Formulation of IT
1. Methanol-isopropanol-water	30	41
2. Acetone-methanol-water	80	105
3. Ethanol-t-butanol-water	79	121
4. C_1-C_5 hydrocarbons	33	32

The results show that the Occam's Razor approach is adequate for thermodynamically ideal mixtures of compounds with small differences in molecular size
while for highly non-ideal mixtures the use of Occam's Razor approach could lead
to severe underdesign and it is essential to use the GMS formulation of IT.
    Similar conclusions in favour of the IT approach to the modelling of
distillation and absorption separations in continuous contacting apparatuses
(e.g. packed and wetted-wall columns) have been reached in other studies carried
out; see Krishna and Taylor [15] for a summary of these findings.
    We had earlier pointed out the need to use the IT formulation to describe
diffusion in the region of the critical point such as the plait point in L-L
systems. Krishna <u>et al</u> [6] have recently reported the results of transient
composition profiles measured in a batch stirred cell with the L-L system:
acetone-glycerol-water. The equilibration trajectory in both liquid phases
(glycerol-rich towards the right and acetone-rich towards the left of Fig. 6)
are highly curvilinear while the Occam's Razor approach would predict equilibration along a <u>linear</u> (dashed in Fig. 6) approach to equilibrium. Use of the
GMS formulation for diffusion in either phase is able to successfully "model"
the experimental results [6]. The L-L behaviour portrayed in Fig. 6 is the
exact analog of the two-bulb gas diffusion experiment of Duncan and Toor [7]
seen earlier; cf. Fig. 3.

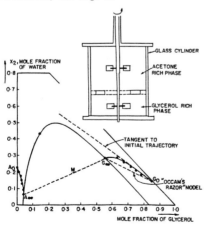

Fig. 6. Composition trajectories in
the L-L system: acetone-glycerol-
water. Results of Krishna <u>et al</u> [6].
It is interesting to observe that
the tangent to the initial traject
in the glycerol-rich phase misses
the binodal curve completely !
Occam's Razor approach is hopelessly
inadequate.

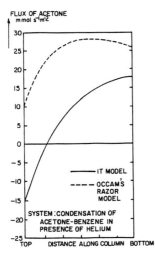

Fig. 7. Condensation of acetone-benzene vapour mixture in the presence of inert gas helium. Experiments carried out in a wetted wall column. For specified inlet conditions at the top of the column, the Fig. 7 shows the acetone fluxes predicted by (1) IT model and (2) Occam's Razor model. Use of Occam's Razor predicts + ve fluxes (i.e. condensation of acetone) all along the column. It was observed experimentally that there was net vaporization of acetone in the column, which happenstance can only be anticipated by the IT model. Results of Krishna [16].

In the foregoing we have demonstrated the utility of using the GMS formulation in describing the interphase mass transfer processes in the widely used separation processes of distillation and extraction. Conventionally used design procedures for these equipment still use the equivalent of the Occam's Razor formulation: Eq. (12), which can be expected to be in serious error in some cases. Let us turn our attention to another important separation process of partial condensation of a vapour mixture.

Separation is achieved when a vapour mixture is cooled below the dew point of the vapour mixture. Industrial condensers are usually operated in a manner that the composition of the liquid condensate is dictated by the transfer fluxes of the components at the point in question:

$$\frac{x_i}{x_j} = \frac{N_i}{N_j} \qquad (23)$$

Also in practice the mass transfer limitations are increased due to the presence of inert gases such as air. Accurate estimation of the transfer fluxes N_i, using the GMS formulation of IT, is required in the design of condensation equipment, as has been emphasized in the review of Krishna and Taylor [15]. In an extreme example of condensation of acetone-benzene in the presence of helium, the work of Krishna [16] has shown that use of a simple-minded Occam's Razor formulation could lead to a wrong anticipation of the <u>direction</u> of transfer of acetone; see Fig. 7. Under the conditions of the experiment described in Fig. 7 reverse diffusion of acetone takes place, a phenomenon earlier signalled during our discussions of the diffusion behaviour of multicomponent gas mixtures.

One reason why the GMS formulation of IT is not used in routine design calculations of separation equipment is possibly due to the (apparent) complexity of the diffusion equations (3)-(4). However, it has been shown by Krishna and Taylor [15] that the GMS diffusion equations can be solved in an efficient manner to yield interfacial transfer fluxes and that these rigorous formulations can be incorporated straightforwardly into efficient computer algorithms for the design of separation equipment. Also, with the use of efficient computational techniques for the solution of the set of design equations (represen ting mass and energy balances, interfacial mass and energy transfer relations and interfacial equilibrium) the IT approach does <u>not</u> require more time than the Occam's Razor approach !

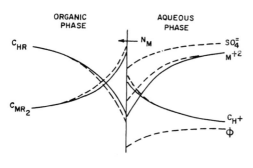

Fig. 8. Concentration profiles during metals extraction:

$$M^{+2} + 2HR = MR_2 + 2H^+$$

in which an aqueous metal ion, such as copper or nickel is exchanged for hydrogen ions by an organic phase solute HR to form an organic soluble metal MR_2. ----- denotes concentration profiles taking electrostatic effects into account; ——— denotes profiles ignoring such effects.

In the foregoing examples of fluid-fluid separation processes, the major, and only, driving force causing relative motion of species (separation) was due to the chemical potential gradients. If the species being separated are charged then the motion of the species will automatically <u>set</u> up an electrostatic potential gradient within each of the fluid phases, which gradient will exert a "pull" or a "push" on the ions depending on the charge and the direction of motion. Neglect of the electrostatic effects can lead to significant errors in the calcultation of the transfer rates, as has been shown by Tunison and Chapman [17] for the case of L-L extraction of metals; see also Fig. 8.   The IT formulation Eqs. (3), (4) and (19) affords a consistent approach of modelling this process.

In L-S separation processes such as crystallization, the knowledge of diffusion coefficients in supersaturated solutions is of importance in the prediction of the rates of crystal growth. Diffusivity data show a very rapid decline in the Fickian diffusivity with increasing concentration in the supersaturated region [18]. This behaviour is analogous to the behaviour noted in L-L systems in the region of the CST [5] and the IT approach is essential for a proper description of the transport process. Compare Figures 2 and 9.

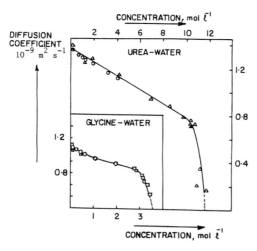

Fig. 9. Diffusion Coefficients in the system urea-water and glycine-water. Data from Myerson and Chang [18].

Let us now turn our attention to another important class of two-phase separation processes in which a fluid is brought into contact with a porous solid;
adsorption and ion exchange fall into this broad category.

Components in both gases and liquids may be separated by selective adsorption on such materials as activated carbon, silica or alumina gel, zeolites etc.
Adsorption can be carried out on both fixed and moving beds of solid adsorbent.
Applications of adsorption include the drying of gaseous streams by use of
silica gel, separation of organic mixtures by adsorption on zeolites, removal of
organic pollutants from waste water streams by adsorption on activated carbon,
water treatment using ion exchange resins, etc.

The most common form of operation is with use of a fixed bed of adsorbent
particles. The fluid feed is made to run through the bed until the bed becomes
nearly saturated and small quantities of adsorbate begins to "break through".
Then the bed must be regenerated to restore its adsorptive capacity and to
recover the adsorbed material. Similarly synthetic ion-exchange resins or some
naturally occurring clays will adsorb ions from aqueous solutions displacing
other ions originally present in the resin matrix until the resin becomes nearly
saturated with the feed stream. Regeneration then follows.

Both adsorption and regeneration are collectively termed "sorption"
processes. Fixed bed sorption processes are fundamentally non-steady state and
the concentrations in the fluid and solid phases inside the bed depend on position and time. When equilibrium between the fluid and solid phases is reached,
the bed loses its sorptive capacity and allows the fluid to pass through the bed
unchanged in composition. The knowledge of the fluid-sorbent equilibria, commonly termed the "sorption isotherm", is a key factor in the choice of a suitable
adsorbent and for design of the bed. The height of the adsorbent bed required
for a specified 'on stream' time is determined by the fluid-solid equilibrium
relationship and by the mass transport processes taking place both outside the
solid phase and in the pores. The modelling of the mass diffusion process on
the outside of the adsorbent is best carried out using the GMS equations (3)-(4)
which reduce to Eqs. (11) for ideal gas mixtures. For transport of mass inside
the pores of the solid some additional mechanisms of mass transport have to be
reckoned with. We first take up the problem of describing mass diffusion of an
ideal gas multicomponent mixture inside porous media and will later extend the
treatment to include diffusion of non-ideal liquid mixtures.

Transport of a gaseous component inside the porous medium is by the following mechanisms:

(i) viscous flow through the pores; this mechanism is non-separative.
It may be noted that this viscous flow mechanism, acting along the direction of
diffusion, is not present when the fluid mixture is not constrained by the walls
of the porous medium, i.e. the diffusion takes place in "open" space.

(ii) Separative, diffusive, transport of the gaseous component through
the pores of the medium. In this case due to the presence of the "inert" wall
of the solid, molecular-wall collisions will occur in addition to molecule-
molecule collisions, the latter only being present for diffusion in open space.
When the diameter of the pore is less than the mean free path of the gaseous
molecule, collisions at the wall "controls" and the mechanism of transport is
termed Knudsen diffusion. On the other hand when the pore diameter is much
greater than the mean free path of the gaseous molecules, the collisions are
mainly between the gas molecules, and "Bulk" gas diffusion mechanism prevails,
as in open space.

Figure 10 shows a schematic of the mechanisms of transport inside the pores
using the analogy with electrical networks; see Jackson [19] and Mason and
Malinauskas [20] for further conceptual discussions. The "total" transfer flux
of component i is the sum of the viscous and diffusive contributions:

$$N_i = N_i^{viscous} + N_i^{diffusive} \tag{24}$$

The viscous contribution to the total flux $N_i^{viscous}$ can be calculated from

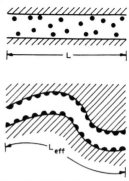

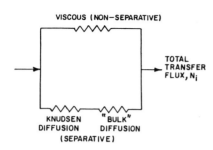

VISCOUS (NON-SEPARATIVE)

TOTAL TRANSFER FLUX, N_i

KNUDSEN   "BULK"
DIFFUSION   DIFFUSION
(SEPARATIVE)

Fig. 10. Electric analogue circuit as a mnemonic device for combining different mechanisms for transport of gaseous species inside porous medium.

Fig. 11. Schematic ways of visualizing the dusty-gas model for gaseous transport inside porous medium [20].

$$\underset{\sim}{N}_i^{viscous} = - x_i \frac{B_0 \, P}{\eta \, RT} \nabla p \tag{25}$$

where B_0 is the permeability of the medium; $B_0 = r_0^2/8$ for a cylindrical pore of radius r_0. The non-separative character of viscous transport is evident from Eq. (25).

Let us now consider the modelling of the diffusive process inside the pores. How do we take the constraints of the walls consistently into account in our IT formulation ? Eq. (11) is valid for ideal gas mixtures diffusing in open space only. An elegant way of extending the IT formulation to include the molecule wall interactions is to consider the wall (porous medium) as the (pseudo) n+1 th component in the mixture, the so-called "dust" molecule [19,20]. These dust species are giant molecules ($M_{n+1} \to \infty$), uniformly distributed in space ($\nabla c_{n+1} = 0$), and are held motionless ($N_{n+1} = 0$) by unspecified external forces acting on them. The precise origin of this external force does not matter in the mathematical treatment; in practice it would usually arise from whatever clamping device holds the porous body stationary. The particular arrangement of the dust particles in space does not matter either, since such geometric characteristics are absorbed into the transport coefficients as the multiplicative factor ε/τ = porosity/tortuosity of the medium. Thus it is unimportant how one chooses to visualize the dust - literally as a random array of large spheres stuck in space, as irregular blobs on the surface of a tortuous capillary, as indicated in Fig. 11, or in some other fashion. See the excellent texts of Jackson [19] and Mason and Malinauskas [20] for a detailed discussion of the Dusty Gas model. One major advantage of the use of the Dusty Gas model is that the Knudsen diffusion coefficient, reflecting the molecule-wall interactions, follows naturally from the kinetic gas theory and is given by:

$$D_{iK}^e = \frac{\varepsilon}{\tau} \frac{2}{3} r_0 \left(\frac{8 \, RT}{\pi \, M_i} \right)^{\frac{1}{2}} \tag{26}$$

which is seen to be independent of pressure in contrast to the GMS molecular diffusivity of the pair i-j, $\mathcal{D}_{ij}$, which is inversely proportional to the pressure. The final working relations for the "total" flux N_i, from both viscous and diffusive contributions is obtained as:

$$\underset{\sim}{d}_i = \nabla x_i + x_i (1 + \frac{B_0 \, P}{\eta \, D_{iK}^e}) \frac{\nabla p}{p} = \sum_{\substack{j=1 \\ j \neq i}}^{n} \frac{x_i \underset{\sim}{N}_j - x_j \underset{\sim}{N}_i}{c_t \, \mathcal{D}_{ij}^e} - \frac{\underset{\sim}{N}_i}{c_t \, D_{iK}^e} \tag{27}$$

The superscript e on the GMS pair diffusivity $Ð_{ij}^e$ serves to emphasize the fact that for bulk diffusion inside the porous medium

$$Ð_{ij}^e = \frac{\varepsilon}{\tau} \, Ð_{ij} \tag{28}$$

Taking note of the fact that for an ideal gas mixture $c_t = p/RT$, the Eq. (27) that Knudsen and Bulk diffusion "regimes" show a different behaviour with regard to the influence of the total system pressure: for operation in the Knudsen regime, the magnitude of the flux N_i will <u>increase</u> with p whereas in the Bulk diffusion regime, N_i remains independent of the system pressure. This point is illustrated by the data presented in Fig. 12, obtained by Mason and co-workers [20], for diffusion of He-Ar in a low permeability graphite septum. At low pressures Knudsen diffusion controls and the fluxes of He and Ar are proportional to the pressure. With increasing system pressure, the mean free path length of the gaseous molecules decreases and at high enough pressures bulk gas diffusion mechanism predominates in which case the fluxes are pressure independent.

Another interesting effect which arises in porous medium gas diffusion is the influence of the pressure gradient, which will act in a "direction" to help the diffusion of one of the species while countering the diffusion process of the other species (acting as a "pull" or a "push"). Figure 13 shows the influence of the pressure difference across the graphite septum on the fluxes of He and Ar. With a positive Δp, the He flux is decreased while the Ar flux is increased in magnitude. The Dusty Gas Model Eq. (27), following IT, models this behaviour properly; see Krishna [21].

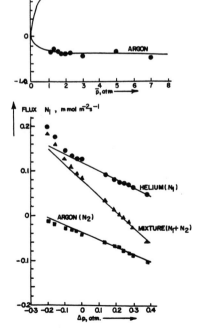

Fig. 12. Influence of system pressure p on the fluxes of He and Ar across a low permeability graphite septum [20]. The experimental results are compared with theoretical predictions of a simplified, linearized, solution to the Dusty Gas Model Eq. (27); see [21].

Fig. 13. Influence of Pressure difference across the graphite septum on the fluxes of He and Ar; experimental data from [20]. The experimental results are compared with the theoretical predictions of a simplified, linearized, solution to the Dusty Gas Model, Eq. (27); see Krishna [21] for details.

The Dusty Gas Model Eq. (27) is most useful in describing the transport rates during separation of gaseous mixtures using porous membranes, as we shall see later. Let us now consider the extension of this approach for treating diffusion of <u>ideal gas</u> multicomponent mixtures in porous media to <u>non-ideal</u> liquid mixtures. The starting point should, needless now to stress, be the GMS Eqs. (3)-(4). By parallel treatment to the Dusty Gas approach, but now including the electrostatic potential gradient term, we obtain (for discussions and derivations see [20,21]):

$$d_i = \frac{x_i}{RT} \nabla_{T,p}\mu_i + \frac{x_i \bar{V}_i}{RT} \nabla p + \frac{x_i B_0}{\eta\, D_{iM}^e} \nabla p + x_i z_i \frac{F}{RT} \nabla\Phi =$$

$$= \sum_{\substack{j=1 \\ j\neq i}}^{n} \frac{x_i N_j - x_j N_i}{c_t\, D_{ij}^e} - \frac{N_i}{c_t\, D_{iM}^e} \tag{29}$$

which equations may be said to be the Dusty Fluid Model Equations [20,21]. Though Eq. (29) is the formal non-ideal liquid analog of Eq. (27), there are some fundamental differences in interpretation of the transport coefficients which must be realised. Firstly, the GMS pair diffusivities D_{ij}^e within the porous medium is no longer simply proportional to the free-space diffusion coefficient D_{ij}, as in the relation (28). In addition the Knudsen diffusion coefficients of the Dusty Gas Model are replaced by D_{iM}^e, the medium or membrane coefficients in order to avoid the connotation of long mean free paths of gaseous molecules. The coefficients D_{ij}^e and D_{iM}^e are parameters to be determined from experimental data but understanding the basis of these coefficients can be expected to aid in the interpretation of observations. The Dusty Fluid formulation Eq. (29) is to be credited to Mason [20] and is, I believe, the most useful form of the rate relations for porous liquid phase mass transport. There are many alternative formulations to be found in the literature [20] and all of these can be shown to be a special case of Eq. (29) above. In particular, the formulation due to Lightfoot [1] is equivalent to Eq. (29) but with the viscous transport term $(x_i B_0 / \eta\, D_{iM}^e)\ \nabla p$ merged into the transport coefficients D_{ij}^e and D_{iM}^e [20].

The electrostatic potential gradient term plays a very important role in the description of the transport phenomena within ion exchange "beads" or particles. On the basis of the previous discussions of transport of ions it should be clear that the rate of transfer of an ion should be very much dependent on the direction in which the ion moves; this directional dependency being caused by the electrostatic "pull" or "push" on the ionic species (cf. Fig. 4). If the ion exchange transfer process is modelled using an "effective" diffusivity without taking account of the electrostatic potential gradient term, then the results can be significantly in error. Figure 14 shows some experimental data which show that loading and regeneration rates occur at significantly different rates [22].

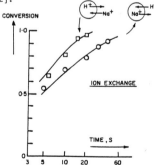

Fig. 14. Exchange rates for H^+ and and Na^+ in an ion exchange column are seen to be strongly dependent on the direction of ion transport; results of Helfferich [22].

6. Membrane Separation Processes

Membrane separation processes, with the exception of osmosis, rely on the differences in the rates of the transfer across the membrane, N_i, to achieve the desired selectivities. There are many ways of classifying membrane processes. From the point of view of understanding the underlying principles, it is best to classify these into two broad categories: (i) porous membranes and (ii) non-porous membranes in which the mechanism of transport across the membrane is "dissolution" of the component(s) in the membrane phase (which could be a swollen polymeric film or a liquid phase stabilized by a surfactant) and transport of the dissolved species across the membrane phase.

6.1 Porous Membrane Separation Processes

Let us consider separation of a gaseous mixture. The transport of the individual species is governed by the Dusty Gas Model Eq. (27). The overall objective is to achieve the desired separation with high selectivities (which means fewer stages) and at lowest possible pressure drop (i.e. low compression costs). How these, conflicting, requirements are satisfied in practice is often a question of compromise. For example, to achieve high selectivities it is best to operate in the Knudsen diffusion regime. An industrially important example is the separation of the isotopes of uranium $U^{235}F_6$ from $U^{238}F_6$ by use of metal barriers which are porous. The separation factor α_{12}, defined by Eq. (15), is

$$\alpha_{12} = (M_2/M_i)^{\frac{1}{2}} \tag{30}$$

which works out to only 1.0043. Many thousand stages, with interstage compression, are required to meet the separation requirements.

Porous membranes have one great advantage over non-porous membranes in that the latter is usually made of a polymeric material and are thus restricted to operating temperatures below about 100 $^{\circ}$C. On the other hand porous membranes can also be manufactured from inorganic materials (ceramics, metals, glass and thus be employed at temperatures of several hundred degrees. Thus porous membranes could conceivably be applied in processes where non-porous membranes fail and the positive separation properties of polymers can no longer be utilized, for example, for the separation of process gases of light molecules which are formed at low pressures and high temperatures; see Eickmann and Werner [23] for further discussions. In particular, the development of a highly porous Al_2O_3 membrane with an average pore radius of 2 to 3 nm by Leenaars et al, cited in [23], appears to make high temperature applications possible.

The desire to obtain high transfer fluxes and high selectivities requires the following membrane conditions (see [23]):
- smallest possible pore diameter of the membrane (less than 100 nm)
- maximum number of pores per unit area
- small membrane thickness
- low system pressure level
- high process temperature

In another type of porous membrane separation process, separation is effected by allowing the components to be separated to diffuse into a third component, called a separating agent, and relying on the differences in the GMS pair diffusivities $Ð_{13}$ and $Ð_{23}$, to obtain the desired selectivities. The separating agent usually used is steam which is later condensed to recover the desired product. The terms Mass and Sweep diffusion are used to describe this process. The components to be separated are made to flow across porous screens, along which we have the flow of teh separating agent. The role of the porous screen is incidental to the process and is <u>not</u> the basis for separation; see Pratt [24] and Cichelli et al [25] for further details. The process is pictured in Fig. 15.

The principle of the mass or sweep diffusion proces can best be understood on the basis of the Dusty Gas Model Eq. (27). We choose a membrane such that bulk diffusion prevails and the pressure gradients are not significantly large. Using subscripts 1,2 to represent the components to be separated and 3 for the sweep species, it can be shown that for the case in which $N_1 = -N_2$, the separation factor α_{12} works out to be

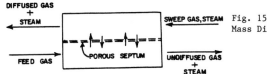

Fig. 15. Schematic of Sweep or Mass Diffusion Process.

$$\alpha_{12} = \exp\left[\frac{N_3\delta}{c_t}\left(\frac{1}{Ð_{23}} - \frac{1}{Ð_{13}}\right)\right] \qquad (31)$$

which shows that large differences in binary pair diffusivities of components 1 and 2 in the sweep gas 3 lead to high separation factors. Practical applications of the sweep diffusion process is the use of steam as separating agent to separate He-Ne, H_2-CO, H_2-Natural Gas mixtures.

Shuck and Toor [26] have demonstrated the use of the sweep diffusion technique for separations involving a liquid mixture of methyl alcohol - n-propyl alcohol - isobutyl alcohol. Equations (29) are the starting point in the estimation of the interfacial transfer rates.

6.2 Non-Porous Membrane Separation Processes

The separation factors obtained using porous membranes, see Eqs. (30) and (31), are limited. Higher selectivities are obtained by use of polymeric membranes or surfactant-stabilized liquid membrane films. The selectivities in non-porous membrane transport arise out of one or more of the following factors:

(i) differences in solubility of the components to be separated in the membrane phase. For example, a mixture of benzene and n-heptane can be separated by use of an aqueous surfactant liquid membrane in which the solubility of the aromatic compound is many hundred times larger than the solubility of the saturated hydrocarbon. Thus by interposing an aqueous layer between the feed mixture and the receiving phase (also a hydrocarbon), selective removal of the aromatic compound from the feed mixture can be achieved. The profile of the overall potential Ψ_i (where $\nabla\Psi_i = d_i$, the overall driving force) for any component i is as shown typically in Fig. 16. At the interfaces I-M and M-II equilibrium is usually assumed to prevail and say at I-M we have (cf. Eq.(22)):

$$\ln\left(\frac{\gamma_{iI}x_{iI}}{\gamma_{iM}x_{iM}}\right) + \frac{(\overline{V}_i)_{avg}}{RT}(p_I - p_M) + \frac{z_i F}{RT}(\Phi_I - \Phi_M) = \Psi_I - \Psi_M = 0 \qquad (32)$$

By choosing the membrane phase M such that the activity coefficient γ_{iM} is large the component i can be effectively "excluded" from the membrane and the desired selectivity can be achieved for the other component(s). The transfer fluxes N_i can be determined from Eqs. (3)-(4) applied to the fluid phases or from Eq. (29) for intra-membrane transport.

In the sub-section (i) under consideration here only the first term of Eq. (32), i.e. the chemical potential term, is relevant.

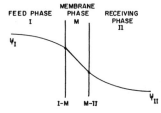

Fig. 16. Typical profiles of overall potential Ψ_i for Membrane separation.

Use of polymeric membranes for gas separations include:
- O_2 enrichment from air (using ethyl cellulose, silicone rubber membrane)
- CO_2 enrichment from air (silicone rubber membrane)
- He separation from natural gas (membrane = cellulose acetate, PTFE, FEP)
- H_2, He recovery from refinery gases (membrane = asym. polysulphone with silicone, polyimide PET)

It needs to be stressed again here that the difference between non-porous and porous membranes is that in the former the permeating species "interacts" with the membrane phase and so the intra-membrane diffusivities of the permeating components have to be determined experimentally. In practice, effective permeability data are measured and reported for a paricular gaseous mixture - membrane system. The Dusty Fluid Model Eq. (29) will help in interpreting the permeability data.

Examples of separation of liquid mixtures using polymeric membranes are: methanol-benzene, benzene-isopropanol, methanol-water, isopropanol-water, ethyl acetate - carbon tetrachloride, o-, m- and p-xylenes, ethanol-acetone; see Rogers et al [27] for a more complete listing of separation possibilities.

(ii) In (i) above the main "driving force" for separation was the chemical potential gradient. In reverse osmosis and ultrafiltration examples, the major driving force is the pressure gradient term. Let us consider an aqueous salt solution separated by a membrane which is permeable only to water and not to the salt. If we further assume that there are no electrostatic potential gradients present, we get from Eq. (22), for water transport

$$\ln\left(\frac{a_{wI}}{a_{wII}}\right) + \frac{\bar{V}_w}{RT}(p_I - p_{II}) = 0 \tag{33}$$

where we have integrated across the membrane (cf. Fig. 16) and a_w represents the activity of water. Since the membrane is permeable only to water the downstream side of the membrane will have only pure water and if we choose the standard state as the one of pure water at pressure p_{II}, it follows that

$$p_I - p_{II} = -\frac{RT}{\bar{V}_w} \ln a_{wI} \tag{34}$$

Since water behaves nearly ideally we may write Eq. (34) as

$$\pi = p_I - p_{II} = \frac{RT}{\bar{V}_w} \ln \frac{p^s_{wII}}{p^s_{wI}} \approx \frac{RT}{\bar{V}_w} \frac{(p^s_{wII} - p^s_{wI})}{p^s_{wI}} \tag{35}$$

where p^s_w represents the vapour pressure of water. The osmotic pressure π, defined as $p_I - p_{II}$ is thus seen to be proportional to the fractional reduction of vapour pressure due to the salt. The constant of proportionality $RT/\bar{V}_w$ has a large value, 138 MN/m^2 at 25 $^\circ C$. This means that even modest vapour pressure lowerings produce a measurable osmotic pressure.

It follows from the above analysis that when an aqueous solution of a salt is separated from pure water by a membrane permeable only to water, water will tend to flow from the region of higher activity (pure water) to the region of lower activity (salt solution) till Eq. (33) is satisfied. The movement of solvent water is called osmosis; this is an equilibrium process and the separation principle can be used for dehydration of food liquids, for example. If a pressure difference exceeding π is applied across the membrane from the side of the salt solution then the water begins to flow from the solution of lower activity to the one of higher activity. This process is reverse osmosis. The retained solute in this case is of the order of < 0.1 nm in size. On the other hand if the retained solute is a macromolecule of the order > 1 nm in size, the process is called ultrafiltration. The mechanism of ultrafiltration is predominantly one of sieving.

The water flux N_w is usually taken to be equal to a permeability times the effective pressure difference: $\Delta p - \pi$.

(iii)   In this third category of non-porous membrane separation processes the electrostatic potential gradient plays a key role, as in electrodialysis. When an ionic solution is subjected to an electric force field then the ions will move according to the flux relation Eq. (19).   A membrane is interposed in the path of diffusion such that only positive ions or negative ions are allowed to pass through the membrane. This ion exclusion from the membrane phase is achieved by incorporating fixed charges on to the polymeric chains making up the membrane. Thus even though the membrane may be physically "porous", the interactions of the diffusing components with the membrane matrix are such that the intramembrane transport must be modelled, using Eq. (29), taking the membrane - solution to be a homogenous phase. If the two types of membranes (allowing +ve and −ve ions respectively) are placed in alternate fashion in a battery, then it is possible to concentrate the cations and anions in one of the compartments in the battery; in the adjacent compartment the ions be in a depleted state. This is the principle of electrodialysis (pressure gradients are usually unimportant here) and the process has been used for the desalination of water, preparing boiler feed water, recovery of brine from sea water, deashing of sugar solution and deacidification of fruit juices. Electrodialysis is particularly useful for separations involving extremely small salt concentrations.

We have discussed above the three main driving forces which are used to effect transfer of a component across a non-porous membrane. The selectivity is achieved by choosing the membrane phase which, generally speaking, "interacts" with the feed mixture in such a way that one or more species in the solution are excluded from the membrane phase, for example by use of electrostatic repulsion (as in electrodialysis). In addition the transfer of the "desired" component to be transported may be enhanced by preferential "complexing" with an "active" chemical species present within the membrane phase. This active species serves to transport the desired material with 100 % selectivity across the membrane in a kind of shuttle service because the original compound gets released at the other end of the membrane phase. The scope for ingenuity in choosing the proper membrane is almost unlimited. As we did in the case of porous membranes, let us list the desirable set of features for non-porous membrane transport:

- highest possible selectivity of membrane
- maximum surface area of membrane in a given volume of module (this is achieved in practice by use of ultra-thin hollow fibre membranes, liquid membranes with tiny microdroplets of about $1 - 5$ μm in diameter, etc)
- small membrane thickness (this is necessary for increasing the flux N_i)
- the membrane must be stable and not have "pinhole defects", "leaks" etc. (such "leakage" will lead to non-selective transport across the membrane and the overall selectivity is thus reduced).

It is interesting to note here that non-selective transport across a non-porous membrane caused by leakage is exactly analogous to the mechanism of non-separative viscous flow which occurs within a porous membrane; see Fig. 10. In a recent paper Krishna and Goswami [28] modelled liquid membrane transport as a parallel step mechanism; see Fig. 17. This model can be extended to non-porous membranes in general.

SELECTIVE TRANSPORT BY DIFFUSION

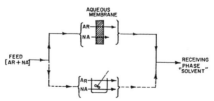

NON-SELECTIVE TRANSPORT BY LEAKAGE

Fig. 17. Separation of Aromatics (A) from Non-Aromatics (NA) using Liquid Membranes [28]. Model taking leakage due to emulsion breakage into account.

7. Concluding Remarks

In this paper we have shown that the Generalized Maxwell-Stefan formulation of the mass diffusion rate relations, based on Irreversible Thermodynamics, Eqs. (3)-(4), provide a general basis for understanding and describing both equilibration and rate governed separation processes.

Specifically, the benefits of adopting a fundamental IT approach have been shown to be the following:

(i)   the IT formulation affords a consistent and correct approach to the description of non-ideal liquid phase transport; in particular the influence of solution thermodynamics on the Fickian diffusivity is made "transparent".

(ii)   the IT formulation is indispensible in describing the transport behaviour in the region of the critical point; this is particularly important in the design of processes involving phase transitions such as crystallization.

(iii) the various driving forces causing relative motion of species, i.e. separation are clearly identified. This identification will aid the screening of alternative separation routes in a systematic manner. In future computer based synthesis of separation schemes can be developed using the IT formulation.

(iv)   the Generalized Maxwell-Stefan Eqs (3)- (4), yield as special cases the widely accepted Maxwell-Stefan equations for ideal multicomponent gas diffusion and the Nernst-Planck equations for dilute electrolyte transport. There is no need for ad hoc modelling of special cases.

(v)   the generalized phase equilibrium relationship Eq. (22), derived from the GMS equations, provides the correct starting point for describing phase equilibrium for equilibrium separation processes.

(vi)   the GMS formulation, can be extended in a logical manner to the treatment of diffusion inside porous media, by modelling the medium as giant "dust" species. For gaseous transport, this approach is particularly rewarding while for liquid phase transport inside porous media, the Dusty Fluid Model will help in the modelling and interpretation of experimental data.

(vi)   The GMS approach has been shown to be absolutely necessary in a few cases involving multicomponent mixtures. Simple minded approaches based on the Occam's Razor approach (i.e. equal transport facility for all components in a fluid mixture) have been shown to lead to significant deviations from experimental observations. In a few extreme cases the Occam's Razor approach has been shown to predict the wrong "direction" of transfer. One separation process, viz. Mass or sweep diffusion relies on the multicomponent diffusion "effects" to achieve separation.

(vii) Most of the commonly used, simplified, flux expressions used for membrane transport processes have been shown to be special cases of the GMS formulation; this leads to a better understanding of the limitations of the simplified approaches.

Finally, it may be expected that with a thorough fundamental approach to the theory of separation processes, the development of "novel" separation techniques could be "systematized".

Symbols Used

a_i	activity of component i in solution, $a_i = \gamma_i x_i$ [-]
a_w	activity of water [-]
B_0	permeability of the porous medium [m^2]
c_i	molar concentration of species i [kmol m^{-3}]
c_t	mixture molar concentration [kmol m^{-3}]
d_i	generalized driving force for the motion of species i relative to mixture [m^{-1}]
$\mathcal{B}_{ij}$	GMS diffusivity of i-j pair in multicomponent mixture [$m^2 s^{-1}$]
$\mathcal{B}$	GMS diffusivity taken equal for all components in the diffusing mixture [$m^2 s^{-1}$]
$\mathcal{B}^e_{ij}$	effective GMS diffusivity of pair i-j in porous medium [$m^2 s^{-1}$]
D^e_{iK}	effective Knudsen diffusion coefficient of gaseous component i in porous medium [$m^2 s^{-1}$]
D^e_{iM}	effective "membrane" diffusivity of component i in porous medium [$m^2 s^{-1}$]
f	rotational speed of centrifuge [s^{-1}]
F	Faraday's constant [9.65×10^7 C kgequiv^{-1}]
G_{ij}	elements of the Hessian matrix [G], with elements given by Eq. (9) [J kmol^{-1}]
[G]	Hessian matrix of the molar Gibbs free energy [J kmol^{-1}]
$\underline{J}_i$	molar diffusion flux of i with respect to molar average mixture velocity [kmol $m^{-2} s^{-1}$]
k_{iT}	thermal diffusion ratio of component i [-]
M_i	molar mass of species [kg kmol^{-1}]
n	number of species in the mixture [-]
$\underline{N}_i$	molar flux of species i in a fixed coordinate reference frame [kmol $m^{-2} s^{-1}$]
$\underline{N}_t$	mixture molar flux in a fixed coordinate reference frame [kmol $m^{-2} s^{-1}$]
p	system pressure [N m^{-2}]
p_w^s	vapour pressure of water [N m^{-2}]
r	radial coordinate [m]
r	radius of ultracentrifuge [m]
r_0	pore radius [m]
R	gas constant [8314.4 J kmol^{-1} K^{-1}]
T	absolute temperature [K]
$\underline{u}_i$	velocity of diffusing species i in a fixed coordinate reference frame [m s^{-1}]
$\underline{u}$	molar average velocity of mixture in fixed coordinate reference frame [m s^{-1}]
$\overline{V}_i$	partial molar volume of species i [m^3 kmol^{-1}]
x_i	mole fraction of species i in "x" phase [-]
y_i	mole fraction of species i in "y" phase [-]
z_i	charge on species i [-]

Greek Letters

α_{ij}	separation factor in a single stage for the pair of components i-j [-]
γ_i	activity coefficient of species i in solution [-]
Γ	thermodynamic correction factor, defined by Eq. (7) [-]
δ	length of diffusion path [m]
ϵ	porosity of medium [-]
η	viscosity of fluid mixture [Pa s]
μ_i	molar chemical potential of species i [J kmol^{-1}]
π	osmotic pressure [N m^{-2}]
ρ_i	mass density of component i [kg m^{-3}]
ρ	mixture mass density [kg m^{-3}]
σ	rate of entropy production [J $m^{-3} s^{-} k^{-1}$]
τ	tortuosity of porous medium [-]
ϕ_i	$= c_i \overline{V}_i$, volume fraction of species i [-]
Φ	electrostatic potential [V]
Ψ_i	generalized potential for mass diffusion; $\nabla \Psi_i = d_i$ [-]

Greek Letters (contd)

ω_i	mass fraction of component i [-]
Ω	angular velocity of rotation of centrifuge [s^{-1}]

Subscripts

avg	averaged over the diffusion path
i,j	referring to species i, j in multicomponent mixture
K	Knudsen diffusion coefficient
M	membrane coefficient
n	referring to nth species in the multicomponent mixture
n+1	referring to the medium which is modelled as a giant "dust" molecule fixed in space
T,p	evaluated at constant temperature and pressure conditions
w	water
1,2,3	referring to species 1,2,3 in ternary mixture
I	referring to phase I
II	referring to phase II

Superscripts

diffusive	diffusive contribution to the transfer flux $\underline{N}_i$
e	effective coefficient inside porous medium
viscous	viscous contribution to the transfer flux $\underline{N}_i$
0	denotes infinite dilution value
—	overbar denotes partial molar value

Vector Notation

$\underline{\nabla}$	gradient operator
$\bullet$	dot product between two vectors

Operators

$\sum$	summation over n species
Δ	difference operator
$\dfrac{d}{dr}$	gradient in r direction

References

[1] E.N. Lightfoot "Transport Phenomena and Living Systems", John Wiley, New York, 1974.

[2] G.L. Standart, R. Taylor and R. Krishna, Chem.Eng.Commun., 3, 277 (1979)

[3] G.J. Dudley and H.J.V. Tyrrel, J.Chem.Soc. Farad Trans. I, 69, 2200 (1973)

[4] A. Vignes, Ind.Eng.Chem. Fundamentals, 5, 189 (1966)

[5] R. Haase and M. Siry, Zeitschrift fur Physikalische Chemie, Neue Folge, 57, 56 (1968)

[5a] M. Modell and R.C. Reid, "Thermodynamics and Its Applications", 2nd Edition, Prentice-Hall, Englewood Cliffs, N.J., 1983

[6] R. Krishna, C.Y. Low, D.M.T. Newsham, C.G. Olivera-Fuentes and G.L. Standart, Chem.Engng Sci., 40, 893 (1985)

[7] J.B. Duncan and H.L. Toor, A.I.Ch.E.J., 8, 38 (1962)

[8] J.R. Vinograd and J.W. McBain, J.Am.Chem.Soc., 63, 2008 (1941)

[9] K. Clusius and G. Dickel, Naturwiss., 26, 546 (1938)

[10] W.M. Rutherford, Separation and Purification Methods, 4, 305 (1975)

[11] J.M. Prausnitz, "Molecular Thermodynamics of Fluid Phase Equilibria", Prentice Hall, N.J., 1969

[12] R.C. Reid, J.M. Prausnitz and T.K. Sherwood, "The Properties of Gases and Liquids", 3rd Edn., McGraw-Hill, New York, 1977

[13] R. Krishnamurthy and R. Taylor, A.I.Ch.E.J., 31, 456 (1985)

[14] R. Krishnamurthy and R. Taylor, A.I.Ch.E.J., 31, 1973 (1985)

[15] R. Krishna and R. Taylor, "Multicomponent Mass Transfer: Theory and Applications", Chapter in Handbook of Heat and Mass Transfer, N.P. Chereminisoff (editor), Vol. 2, Gulf Publishing Corp., Houston, 1986

[16] R. Krishna, Trans.Inst.Chem.Engrs, 59, 35 (1981)

[17] M.E. Tunison and T.W. Chapman, Ind.Eng.Chem. Fundamentals, 15, 196 (1976)

[18] A.S. Myerson and Y.C. Chang, "Diffusion Coefficients in Supersaturated Solutions", Ms in press; see also Y.C. Chang, Ph.D. Thesis, Georgia Institute of Technology, Atlanta, Georgia (1984)

[19] R. Jackson, "Transport in Porous Catalysts", Elsevier, Amsterdam, 1977

[20] E.A. Mason and A.P. Malinauskas, "Gas Transport in Porous Media: The Dusty Gas Model", Elsevier, Amsterdam, 1983

[21] R. Krishna, "A Simplified Procedure for the Solution of the Dusty Gas Model Equations for Steady-State Transport in Non-Reacting Systems", Chem.Eng.Journal, in press

[22] F. Helfferich, J.Phys.Chem., 66, 39 (1962)

[23] U. Eickmann and U. Werner, German Chemical Engineering, 8, 186 (1985)

[24] H.R.C. Pratt, "Countercurrent Separation Processes", Elsevier, Amsterdam, 1967

[25] M.T. Cichelli, W.D. Weatherford Jr and J.R. Bowman, Chem.Eng.Progr., 47, 63 (1951); Ibid, 47, 123 (1951)

[26] F.O. Shuck and H.L. Toor, A.I.Ch.E.J., 9, 442 (1963)

[27] C.E. Rogers, M. Fels and N.N. Li, in "Recent Developments in Separation Science", N.N. Li (editor), The Chemical Rubber Company, Ohio, 1972

[28] R. Krishna and A.N. Goswami, "Influence of Emulsion Breakage on Selectivity in the Separation of Benzene-Heptane Mixtures using Aqueous Surfactant Membranes", to be published in the Proceedings of the International Solvent Extraction Conference, Munich, September 1986.

The Transport of Water in Polymers

By J. A. Barrie

DEPARTMENT OF CHEMISTRY, IMPERIAL COLLEGE OF SCIENCE AND TECHNOLOGY, LONDON SW7 2AY, UK

The sorption and diffusion of water in polymers is of importance in a number of areas of technological and biological interest and a variety of polymers, both synthetic and natural, have been studied. An important feature of these systems is the ability of the water molecule to form hydrogen bonds with other water molecules and/or with polar groups in the polymer. Other substances such as alcohols and phenols also have this ability to associate or interact through the formation of hydrogen bonds. In the context of gas-mixture separations using polymer membranes, many gas streams of commercial importance contain water vapour which may affect both gas permeability and the effectiveness of the membrane as a permselectivity barrier. Moreover, the application of membrane separations to the drying of wet gas streams is itself of interest.

There is an extensive literature on the sorption and diffusion of water in polymer films which has been the subject of a number of reviews.[1-6] In the present paper a comparatively brief description is given of a few selected features which may be relevant to membrane separations.

Types of Sorption and Transport Behaviour

A rough subdivision of water-polymer systems into two or three groups can be made on the basis of the intrinsic

hydrophilic or hydrophobic character of the polymer which will
determine to some extent the physical state of the sorbed water.

The first group comprises materials with saturation regains
of around 5wt% or more and includes the natural fibres, proteins,
polyamides, polyimides, epoxies and water soluble polymers.
Most of these contain polar groups with which the water molecule
can form hydrogen bonds. The equilibrium regain will reflect
to some degree the strength of interchain bonds and their suscept-
ibility to attack by water. In semicrystalline polymers the
sorption of water is restricted to the amorphous phase and
to the surface of crystallites. The following features are
characteristic of this group and are illustrated in Figures 1 to 4.

There is often a Langmuir-type curvature to the isotherm
at low activities generally indicative of a competitive
sorption on a fixed number of sites and the isotherm shape
is type II on the BET classification. The presence of
polar groups with the ability to form hydrogen bonds does
not, in itself, guarantee high levels of sorption as demonstrated by the relatively low values for the polyamides[1,7-11]
and phenoxy polymer.[12] With the water soluble polyacrylamide
the more detailed study indicates a Langmuir-type knee[13,14]
whilst for polyvinylalcohol, sorption at lower activities
is substantial but the knee is weak for high crystallinity
samples.[15,16] It is probable that disruption of crystal-
lities or network structure occurs with polyvinylalcohol
at an early stage to produce additional sites for water
sorption. By comparison the level of sorption for
polyethyleneoxide, which is also water soluble, is much
less at the lower activities; complex formation has been
suggested as a basis for the favourable interaction at

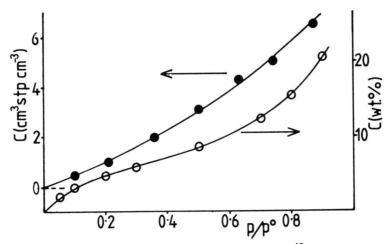

Fig. 1 Sorption isotherms for water at 50°C: ○, cellophane[42] and ●, polybutylmethacrylate.[22] $1 cm^3 stp/cm^3 = 8.03 \times 10^{-2}/\rho$ wt% where ρ is the density of the polymer.

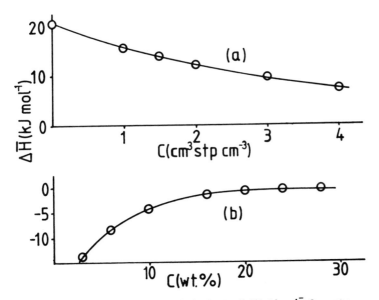

Fig. 2 Concentration dependence of the heat of dilution $\Delta \bar{H}$ for water: (a) polybutylmethacrylate[22] and (b) cellophane.[42]

higher activities.[5] There have been many attempts to rationalize the affinity of a polymer for water. In general this is not simply a function of the concentration of polar groups but will depend also on their nature and position on the polymer chain as well as factors such as degree of crystallinity.

The heat of dilution at low regains is often negative and tends to zero as the activity approaches unity consistent with the water sorbed initially being more strongly bound than in the liquid state. Support for at least two physical states for the sorbed water, in one of which the molecular motion is more restricted, is claimed from a number of investigations using calorimetric and spectroscopic techniques but there are differences of interpretation.[17,18,19]

The steady-state flux generally increases non-linearly with the activity such that the permeability increases with concentration especially at the higher regains. There is often a significant reduction in glass transition temperature accompanying the sorption, as, for example, with polyvinylalcohol and nylon 6,[20] suggesting that network plasticization is responsible for the concentration dependence of the permeability.

The diffusion coefficient usually increases and the corresponding activation energy decreases with concentration. At the higher activities this behaviour is again consistent with plasticization which promotes the diffusion process. Substantial swelling of the network may occur and pronounced non-Fickian sorption kinetics are encountered when the system is close to or below the glass transition temperature.

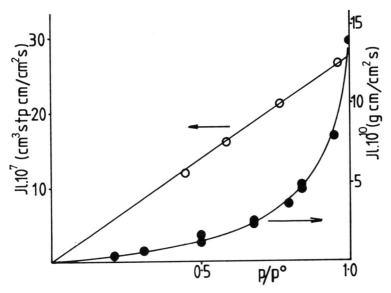

Fig. 3 Dependence of the steady-state flux J on water vapour activity at
40°C: ○, polybutylmethacrylate,[22] ι = 0.01cm and ●, nylon film,[94]
ι = 0.003cm.

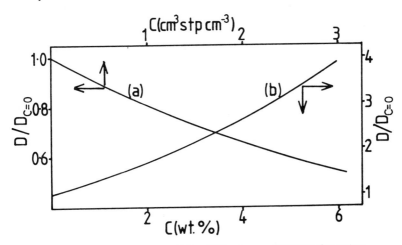

Fig. 4 Concentration dependence of the diffusion coefficient for water
at 40°C: (a) polybutylmethacrylate,[22] $D_{C=0} = 6.7 \times 10^{-7} cm^2 s^{-1}$,
(b) nylon,[94] $D_{C=0} = 1.4 \times 10^{-7} cm^2 s^{-1}$.

The behaviour resembles in a number of aspects that of organic vapours in glassy polymer films.

The second group of polymers comprises most elastomers and common engineering plastics with saturation regain less than ~5wt%. The characteristic features are also illustrated in Figures 1 to 4 and are:

Sorption isotherms are generally BET type III and saturation regains as low as 0.1wt% are encountered as for polydimethylsiloxane[21] and polyethylene.[1] Any type II curvature at low activities is sufficiently weak to be difficult to detect.

The heat of dilution is often close to zero over most of the activity range with a tendency to become positive at the lower concentrations. This behaviour is consistent with, although does not necessarily imply, a tendency for the sorbed water to cluster as the regain increases.

The steady state flux is directly proportional to the activity drop across the membrane and the permeability is constant over the whole, or most, ofthe activity range.

The diffusion coefficient decreases and the corresponding activation energy increases with concentration over much of the range.

Many of these features can be discussed in terms of different physical states for the sorbed water as, for example, a dual-state model with a mobile component and a relatively immobile component. The former may be identified with a dissolution component to the sorption and the latter with a Langmuir-type

component or associated water clusters. Assuming that only
the dissolved monomeric water effectively diffuses and that
local equilibrium is maintained between the two species, one
can write, using subscript 1 to denote monomer,[1,3,21,22]

$$J = J_1 = -D_1 \frac{\partial C_1}{\partial x} = -D \frac{\partial C}{\partial x} \qquad (1)$$

and

$$D = D_1 \frac{\partial C_1}{\partial C} \qquad (2)$$

For sufficiently low concentrations of dissolved monomeric
water, $C_1 = k_1 p$, where k_1 is the Henry-law dissolution constant
and D_1 can be taken as constant so that

$$J = -D_1 k_1 \frac{\partial p}{\partial x} \qquad (3)$$

The flux is then proportional to the pressure drop across
the membrane. For immobilization according to the Langmuir
model, the diffusion coefficient according to equ (2) increases
with concentration and in the limit $C \rightarrow 0$ reduces to

$$D_{C=0} = D_1 k_1 / k \qquad (4)$$

where k is the overall Henry-law constant. For systems with
a pronounced Langmuir knee to the isotherm, the ratio k_1/k
will be small and the effective limiting diffusion coefficient
correspondingly smaller than that for the mobile component.

For immobilization through clustering of the sorbed water
both $\partial C_1/\partial C$ and the fraction of mobile water, C_1/C, decrease
with concentration with a limiting value of 1 as $C \rightarrow 0$. It
follows from equ (2) that the effective diffusion coefficient
decreases with concentration and that D_1 may be identified
with $D_{C=0}$ and k_1 with $k_{C=0}$ provided C_1 is sufficiently small.

The permeability $\bar{P} = D_1 k_1 = D_{C=0} k_{C=0}$ is then independent of
concentration. Dimerization or polycondensation of the water
and dissolution of water-soluble impurities are among the models
for clustering which are more amenable to quantitative development.

In the poly condensation model the polymer is regarded
as an inert medium and the water molecule as a tetrafunctional
monomer polymerizing through hydrogen-bond formation. The
gel point of the polycondensation is identified with the
saturation point of the system at $p/p^o = 1$ and an expression
is obtained for C_1 as a function of C. The concentration
dependence of D, E_D and ΔH is predicted but agreement with
experiment is semiquantitative. The increase in E_D with
concentration is attributed to the breakage of hydrogen bonds
as the concentration of monomer increases with temperature
for constant C. To treat the polymer as an inert medium and
to neglect co-operative effects in hydrogen-bond formation
are clearly oversimplifications.

The Zimm-Lundberg analysis which relates the statistical
thermodynamic cluster integral G_{11} has also been widely used
to support the concept of cluster formation. In general,
relatively small clusters are envisaged containing at most
a few molecules, too small to scatter visible-light except
perhaps at activities approaching unity.[23,24]

The introduction of hydrophilic impurities, such as NaCl,
into the polymer network can have a pronounced effect on cluster
size with pockets or droplets of salt solution forming once
the activity exceeds that for the saturated salt solution.
This effect has been demonstrated for elastomers and a quant-
ative development achieved by balancing the osmotic pressure

of the droplet with the opposing network pressure.[25,26] Large additional amounts of water are taken up and the diffusion coefficient may decrease by several orders of magnitude through the term $\partial C_1/\partial C$ in equ. (2). By comparison the flux is proportional to the activity and the permeability is virtually constant and equal to that for the unfilled rubber consistent with the view that only the dissolved monomeric water is responsible for the flux.[27] The bulk of the additional water is effectively immobilized in the droplets and contributes little to the flux or to network plasticization. Related behaviour is found with ionomers, the sorbed water being attracted preferentially to the ionic domains.[28] In some systems the solubility of the metal salt in the polymer phase may be appreciable through strong salt-polymer interactions favouring possible complex formation. A copolymer of ethylene and vinyl alcohol with LiCl additive is a case in point where there is evidence that the bound component of the salt tends to decrease the co-ordination sites available for water sorption.[29-31]

It may be argued that the transport behaviour of this group of materials can be attributed to traces of such hydrophilic impurities as only relatively small amounts are required to alter significantly the sorptive capacity of the polymer for water. However, diffusion coefficients which decrease with concentration are also observed for alcohols and phenols in silicone and natural rubbers and an impurity mechanism is less likely.[32-35] It is probable that both impurity and clustering mechanisms are operative to varying degrees in water-polymer systems.

As indicated earlier this subdivision of behaviour into

groups is by no means strict and is mainly for illustrative
purposes. Thus for some polymers such as polyvinylacetate[1,33]
and some polyurethane elastomers[36,37] the flux increases non-
linearly with the activity such that the permeability increases,
although rather weakly with concentration. The isotherm,
however, is BET type III, any Langmuir-type knee being absent
or sufficiently weak to require the use of sensitive pressure
gauges for its detection. The concentration dependence of
the diffusion coefficient is generally weak and it may be
effectively constant over much of the range. This and related
behaviour may be the result of opposing factors. For example,
a concentration of monomeric water sufficiently high to
plasticize the network will be reflected in D_1 increasing
with C so as to oppose any decrease in $\partial C_1/\partial C$ associated with
clustering.

The isotherm shape may also be affected by the physical
state of the polymer. For polyvinylacetate above the glass
transition temperature the isotherm is BET type III whereas
below the transition it becomes increasingly type II in
character;[38] in this instance microvoids in the glassy state
may act as preferential sites for the initially sorbed water.
Different isotherm types have also been obtained on materials
of apparently identical composition;[39,40] again the microvoid
content of the glass and the influence of thermal and solvent
histories may be contributing factors.

Current gas separator units are based on polyethersulphone
or cellulose acetate membranes. Relatively few studies of
water sorption and transport have been reported for the poly-
ethersulphones or polysulphones;[10,12,41] the sorption isotherm
is linear or shows weak BET type III curvature with saturation

regains approaching ~5wt%. On the other hand, cellulose
acetate with saturation regains in the region of 15-20wt% has
been studied extensively; there is apparent disagreement in a
number of findings.[42-49]

The sorption isotherm for cellulose acetate has been
reported as BET type III but more detailed mapping indicates
type II; a hysteresis loop is also observed on desorption and
dimensional changes accompany the sorption over most of
the activity range.[42,45] Spectroscopic studies indicate that
the average hydrogen-bond strength is lower than for liquid
water for sorption in the lower half of the activity range.
At higher activities it is stronger but there is some disagreement as to whether it compares with bulk water. It is concluded
that the water in cellulose acetate is initially monomolecularly
dispersed with the development of clusters at higher activities;
from infra-red studies clusters of a few molecules are envisaged
whilst larger domains with properties approaching bulk water
are implied from Raman investigations.[50,51]

Heats of dilution are negative and range from around -10
to -4 kJmol^{-1} as the activity is increased suggesting that
the sorbed water and especially that initially sorbed interacts
more strongly with the polymer than with the bulk liquid contrary
to the spectroscopic evidence.[42] The reason for this apparent
discrepancy is not clear but one might speculate that because
of the relatively bulky nature of the cellulose chain there
are numerous small microvoids available for the initial sorption
of the water. The heat of dilution will then comprise terms
from the van der Waal interaction of the sorbate and cavity
as well as any hydrogen bond formation between the water and
polymer. A zero or negative heat of dilution may therefore

be consistent with a relatively weaker degree of hydrogen bonding.
In addition the water sorbed initially in the intersitial spaces
may then act as further sites for water uptake with progressive
disruption of the network.

There is also disagreement in the reported concentration
dependence of the diffusion coefficient. Detailed study indicates
that the system is complex with a number of anomalous non-Fickian
features. Both the solubility and the permeation rate have a long-
term time dependence suggesting slow relaxation and plasticization
effects. A steady state analysis yields a diffusion coefficient
and permeability which is an increasing, albeit not strong,
function of C.[45,46]

Another important aspect of the water-polymer interaction is
its effect on the glass transition temperature[20] and various
mechanical, optical and electrical properties and in the promotion
of crazing and whitening. Many of these aspects have been
reviewed recently.[52,53]

Water Permeability and Ideal Separation Factors

Permeabilities for water vapour in a number of polymers are
given in Table 1 and are expressed as ccstp, cm/cm^2s cmHg. Unless
stated otherwise, these refer to the limit C→O and as such are
lower limits when $\bar{P}$ increases markedly with vapour activity as for
nylon, regenerated cellulose and polyvinylalcohol. In a few cases
estimates of $\bar{P}$ were obtained from $\bar{P} = \bar{D}\bar{S}$ using average diffusion
coefficients from sorption kinetics. Also included in Table 1
are the saturation regains; in a number of cases these were
obtained by extrapolation of isotherm data and as such are
approximate.

Silicone rubber, the cellulose derivatives, the polyether-
based polyurethanes and the urethane-amino a̅c̅ copolymer emerge

TABLE 1: Water Permeability $\bar{P}_{C=0}$ and Saturation Regain C_s

Polymers	T(°C)	C_s(wt%)	$\bar{P}_{C=0} \cdot 10^9$	Ref.
polydimethylsiloxane	30	0.1	4500	21
polyurethane (Adiprene CM)	35	3.5	3200	36
urethane-amino ac. copolymer	38	-	1700 (0.9 p/p°)	55
silicone-polycarbonate copolymer	-	-	100	54
cellulose acetate	30	18	1000	42, 43
cellulose triacetate	30	11	1600	42, 76
ethyl cellulose	30	7.5	2000	40
polyimide (Kapton H)	30	4.2	64	77
bismaleimide (Compimide M751)	40	7	23*	78
epoxy MY720	40	5	14*	70
polymethylmethacrylate	40	2.2	140	22
polyethylmethacrylate	30	1.3	510	22
polyvinylacetate	35	5.8	550	33
polyethersulphone	30	2.2	40*	12
polyethyleneterephthalate	30	0.9	19	77
polyvinylchloride	25	1.0	28	79
polyacrylamide	30	∞	400+	13
poly(γ-methyl-L-glutamate)	20	5.0	800+ (0.4 p/p°)	80
regenerated cellulose	25	40	190	1
polyvinylalcohol	25	-	1.9(0.4 p/p°)	1
polyacrylonitrile	30	5.6	3.8	83, 84
poly(monocloro-p-xylene)	30	0.2	1.9	77
nylon-6	25	10	40 (0.5 p/p°)	82
polycarbonate	25	0.38	110	85
polystyrene	25	0.07	97	86, 1
natural rubber	30	-	270	32
poly(2,6dimethyl phenylene oxide)	30	-	406	82
phenoxy	30	3.3	-	81

* Estimated from $\bar{P} = \bar{D}\bar{S}$

+ Depends on casting solvent

as the high permeability materials. However, $\bar{P}$ for nylon, regenerated cellulose and polyvinylalcohol will increase by an order of magnitude or more as the vapour activity approaches unity. A high permeability may be the result of a high diffusion coefficient, a high solubility or moderately high values of both; for example, $D_{C=O}$ for silicone rubber is several orders of magnitude greater than that for cellulose acetate.

Water permeabilities are generally higher than those for gases largely because of the higher solubility for water. In Table II permeabilities for carbon dioxide and the ideal separation factor $\alpha(A,B) = \bar{P}(A)/\bar{P}(B)$ are presented. Based simply on the data of Tables I and II a number of materials appear to have the potential for the drying of wet gas streams of carbon dioxide and methane. For an ideal system of non-interacting fluxes and given membrane configuration $Q(CH_4)/Q(H_2O)$ = $\bar{P}(CH_4).p(CH_4)/\bar{P}(H_2O).p(H_2O)$ and a favourable permeability ratio may be offset appreciably by a less favourable ratio of the partial pressures of the components. Ideally for the removal of both water and carbon dioxide a high $\bar{P}(H_2O)$ coupled with a low $\alpha(H_2O,CO_2)$ and a high $\alpha(CO_2,CH_4)$ is preferred. Clearly some compromise has to be made and as well as the more conventional cellulose acetate, polysulphone and silicone-polycarbonate block copolymers others such as polyvinylacetate, polyphenylene oxide, polyurethanes, regenerated cellulose and Kapton are also worthy of consideration as potential membrane separators. In practice many other factors must be considered, including membrane geometry and effective thickness, and permeator design factors such as single or multistage operation and recycling.[54]

TABLE 2: Carbon Dioxide Permeabilities and Ideal Separation Factors

Polymer	$T^{\circ}C$	$\bar{P}_{CO_2} \cdot 10^9$	$\alpha(H_2O,CO_2)$	$\alpha(CO_2,CH_4)$	Ref.
polydimethylsiloxane	25	270	17	3.4	87
silicone/polycarbonate	30	8.4	12	19.0	54
natural rubber	25	13.0	21	4.3	88
polyimide (Kapton)	35	0.03	2133	64	92
polysulphone	35	0.65	-	28	92
polycarbonate	35	0.85	130	24	92
poly(2,6 dimethyl phenylene oxide)	35	7.8	52	15	92
cellulose acetate	35	0.6	1667	31	92
polyvinylacetate	35	0.22	2500		95
polyethyleneterephthalate	30	0.014	1357		96
polyvinylchloride	25	0.016	1750		91
polystyrene	35	1.1	$^{\sim}88$		93
polyethylmethacrylate	25	$1.8 \cdot 10^{-4}$*	1020		91
polyacrylonitrile	30	0.44	21000	39	82
polyethersulphone	30	0.016	91		89
nylon 6	30	1.10^{-3}	2500		82
polyvinylalcohol	23	$0.5 \cdot 10^{-3}$	19000		90
cellophane	25	13.7	380000		58
ethylcellulose	30		146		67
urethane-amino ac copolymer	38	1.7 (CH_4)	$1000{:}(\alpha(H_2O,CH_4))$		55

Brief mention is made of the scope for widening the choice
of membrane material through techniques such as copolymerization,
chemical modification and blending which, in principle allow
for a degree of control over the water permeability. The
polyurethane-amino $\overline{ac}$[55] and silicone-polycarbonate[54] are examples
of random and alternating block copolymers respectively; others
such as ethylene-vinylalcohol copolymers which combine hydrophobic
and hydrophilic structural units are of interest but there
are few, if any, detailed studies of water transport in this
material. The polyurethanes afford examples of chemical modification through variation of the polyether content or introduction
of charged or polar groups. Blending of polymers may be used
to improve processing or mechanical properties as well as
control the permeability. Blends may be immiscible (heterogeneous), miscible (homogeneous) or partially miscible
(heterogeneous) and include graft copolymers. The literature
on gas, vapour transport particularly in heterogeneous blends
is significant but studies with water are comparatively few
especially for miscible blends.

Interaction of Gas and Water Fluxes

That the presence of water vapour may accelerate the
sorption rates of vapours in polymers is well known and is
attributed to the relatively rapid diffusion of the water
and its plasticizing action on the network.[56] With gas, water
vapour mixtures, the ideal situation with the respective fluxes
non-interacting is only likely to obtain with the more hydrophobic materials such as polyethylene and silicone rubber.
With the more hydrophilic materials such as regenerated cellulose,

polyvinylalcohol and nylon the gas permeability increases
rapidly with vapour activity especially at higher regains
when plastization of the networks is most evident. For cellophane
increases in gas permeability of several hundred or more are
observed, most of which occurs in the region of activity greater
than 0.6.[57-62]

More recent work on regenerated cellulose confirms the
earlier measurements and extends them to other gases on both
branches of the hysteresis loop. For regains of less than
10%wt of water, the permeability of hydrogen decreased and
passed through a minimum before showing the usual increase
associated with plasticization as shown in Figure 5. It was
also observed that $\bar{P}_{H_2}$ for a given regain was greater on the
desorption side of the loop and was associated with the higher
volume of the sample on the desorption side.[63,64] For feed
mixtures of CO_2 and CH_4 permeating membranes of cellulose
acetate, polyethersulphone and polysulphone the combined gas
flux passed through a maximum as the water content of feed
gas was increased and in all cases there was a significant
reduction in the gas separation factor.[65]

For a number of polymers reductions in the gas permeability
are observed in the presence of water vapour. For H_2, CH_4
mixtures at 50% humidity and 30°C reductions of up to 60 per
cent were recorded in the permeabilities of both components;
the reduction in selectivity was much less. At 100°C the
water vapour had no noticeable effect on the gas permeabilities.
It was postulated that the water preferentially sorbs in and
excludes the gas from the microvoid content of the network
reducing the available diffusion pathways for the gas.[66]

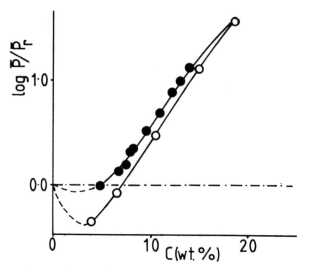

Fig. 5 Effect of Water Regain on the relative permeability of regenerated
cellulose to hydrogen at 25°C.[64] ○, sorption branch of hysteresis
loop; ●, desorption branch. $\bar{P}_r$ is the hydrogen permeability in the dry
film.

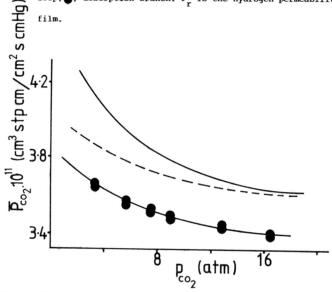

Fig. 6 Effect of Water Vapour on the permeability of Kapton to carbon dioxide
at 60°C.[97] ●, 9.3% relative humidity in the feed stream; (—), dry carbon
dioxide; (---), model assuming simple dual mode competition; (-●-),
model accounting for anti-plasticization of Kapton by H_2O.

The permeability of carbon dioxide in ethyl cellulose also
decreased in the presence of water vapour, the effect being
more marked when both membrane faces, rather than the upstream
only, were exposed to the water vapour.[67] In a more recent
study a similar depression of the carbon dioxide flux through
Kapton, as illustrated in Figure 6, has been analysed in terms
of the dual mode theory of sorption and transport in glassy
polymers. The decrease in permeability was greatest at low
water contents where population of the microvoid content occurs
to the exclusion of the gas; for this system it is also argued
that the water induces a degree of antiplasticization so as
to reduce by a small amount the diffusion coefficient of the
dissolved species.[68]

The interaction of water with glassy polymers on time
scales which are long relative to the time to eliminate concentration gradients by pure diffusional processes is complex.
In common with organic vapours the sorption and transport
of water vapour is a function of solvent and thermal history.
Both reversible and irreversible components may contribute
to swelling of the matrix on absorption of vapour and on subsequent desorption a frozen-in expanded structure obtains.[69-71]
This behaviour is closely related to the anomalous or non-
Fickian diffusion of vapours in glassy polymers and similar
behaviour is observed with the more strongly sorbed gases
such as carbon dioxide.[72-75] The increased sorption of the
expanded structure obtained by ageing is illustrated in Figure 7
and 8 for epoxy-water systems.

In conclusion it is pertinent to examine what is implied
by plasticization in this context. One may distinguish between

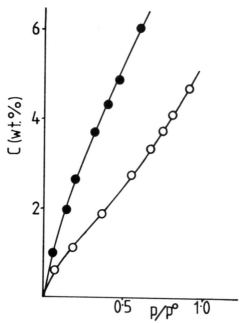

Fig. 7   Sorption isotherms for water at 40°C in epoxy resin.[98] O -"first"
sorption;   ● - after ageing at 150°C in water for 7 days.

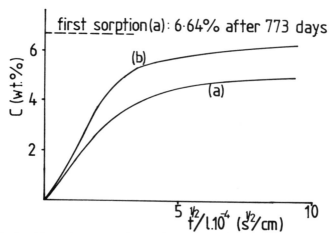

Fig. 8   Sorption rate curves at room temperature for water/epoxy system.[99]
(a) first sorption and (b) after the first sorption (773 days) the
sample was dried in vacuum oven at 100°C prior to second sorption.

systems where the sorption of water reduces the glass transition temperature below that of the experiment and those for which the reduced system T_g is well above that of the experiment as is the case of the epoxies and polyimides. The first case is normal plasticization of the network accompanied by gross changes in segmental mobilities which may facilitate transport of other components. In the second case the system may be well below its T_g and the dissolved water induces network relaxations on a long time scale which lead to a gradual expansion of the network. Although plasticization of backbone motions on the long time scale is evident, the effect on a diffusional time scale may be practically negligible and the transport of gaseous component little affected by this mode even though expansion of the network has occured. As the time scale of relaxation shortens the normal plasticization mode will become more effective.

REFERENCES

1. J A Barrie, In "Diffusion in Polymers" (J Crank and G S Park eds). Academic Press, London, 1968, Chapter 8.

2. D Machin and C E Rogers, CRC Critical Reviews in Macromolecular Science, CRC Press, Cleveland, 1972.

3. D Machin and C E Rogers, Encyclopedia of Polym. Sci. and Tech., 1970, 12, 679.

4. P Meares, Philos. Trans. R. Soc. London Ser. B. 1977, 278, 113.

5. P Molyneux, In "Water" Vol.4 (F Franks ed) Plenum Press, London, 1975, 569.

6. V Stannett, H B Hopfenberg and J H Petropoulos. In "Macromolecular Science, Vol.8, Ed. C E H Bawn (MTP International Review of Science) Butterworths, London, 1972, 329.

7. R Puffr and J Sebenda, J. Polym. Sci. (C), 1967, 16, 79.

8. A Sfirakis and C E Rogers, <u>Polym. Eng. and Sci.</u>, 1980, <u>20</u>, 294.

10. G Skirrow and K R Young, <u>Polymer</u>, 1974, <u>15</u>, 771.

11. K Inone and S Hoshino, <u>J. Polym. Sci. Polym. Phys. Ed.</u> 1976, <u>14</u>, 1513.

12. V B Singh, J A Barrie and D J Walsh, <u>J. Appl. Polym. Sci.</u> 1986, <u>31</u>, 295.

13. A Kishimoto and T Kitahara, <u>J. Polym. Sci. (A1),</u> 1967, <u>5</u>, 2147.

14. J C Chuang and H Morawetz, <u>Macromolecules</u>, 1973, <u>6</u> 43.

15. A Takizawa, T Negishi and K Ishikawa, <u>J. Polym. Sci. (A1)</u>, 1968, <u>6</u>, 175.

16. A W Myers, J A Meyer, C E Rogers,V Stannett and M Szwarc, <u>Tappi</u>, 1961, <u>44</u>, 58.

17. K Packer, <u>Philos Trans. R. Soc. London Ser. B.</u> 1977, <u>278</u>, 59.

18. G E Johnson, H E Bair, S Matsuoka, E W anderson and J E Scott, In "Water in Polymers" ACS Symp. Ser. 127, 1979, 451.

19. C A J Hoeve, In "Water in Polymers" ACS Symp. Ser. 127, 1979, 135.

20. H K Reimschuessel, <u>J. Polym. Sci., Polym. Phys. Ed.</u>, 1978, <u>16</u>, 1229.

21. J A Barrie and D Machin, <u>J. Macromol. Sci. Phys.</u>, 1969, <u>33</u>, 645.

22. J A Barrie and D Machin, <u>Trans. Faraday Soc.</u> 1971, <u>67</u>, 245.

23. J L Lundberg, <u>Pure and Appl. Chem</u>, 1972, <u>31</u>, 261.

24. A Misra, D J David, J A Snelgrove and G Matis, <u>J. Appl. Polym Sci.</u>, 1986, 375.

25. E Southern and A G Thomas, In "Water in Polymers", ACS Symp. Ser. 127, 1979, 375.

26. K Muniandy and A G Thomas, <u>Trans I Mar. E(C) 97 Conf. 2</u>, 1985 87.

27. J A Barrie and D Machin, <u>J Macromol Sci. Phys.</u>, 1969, <u>B3</u>,673.

28. D Graiver, M Litt and E Baer, <u>J. Polym. Sci. Polym. Chem. Ed.</u> 1979, <u>17</u>, 3589.

29. S Gaeta, A Apicella and H B Hopfenberg, <u>J. Membrane Sci.</u>, 1982, <u>12</u>, 195.

30. H B Hopfenberg, A Apicella and D E Saleeby, J. Membrane Sci.,
    1981, 8, 273.

31. A Apicella and H B Hopfenberg, J. Appl. Polym. Sci., 1982,
    27, 1139.

32. J A Barrie, D Machin and A Nunn, Polymer, 1975, 16, 811.

33. A C Sheer, Thesis (University of London), 1972.

34. R S Smith, Thesis (University of London), 1972.

35. J A Barrie, J. Polym. Sci. (A1), 1966, 4, 3081.

36. J A Barrie, A Nunn and A Sheer, In "Permeability of Plastic
    Films and Coatings" (H B Hopfenberg ed.), Plenum Press,
    London 1974.

37. N S Schneider, L V Dusablon, L A Spano, H B Hopfenberg and
    F Votta, J. Appl. Polym. Sci. 1968, 12, 527.

38. Z Miyagi and K Tanaka, Colloid and Polym. Sci., 1979, 257,
    259.

39. H Yasuda and V Stannett, J. Polym. Sci. 1962, 57, 907.

40. J D Wellons and V Stannett, J. Polym. Sci. (A1) 1966, 4, 593.

41. E Cole and L S A Smith, Trans. I. Mar. E (C) 97 Conf. 2., 1985,
    149.

42. R Jeffries, J Textile Institute, 1960, 51, T339, T399, T441.

43. J D Wellons, J L Williams and V Stannett, J. Polym. Sci. (A1)
    1967, 5 1341.

44. H B Hopfenberg, F Kimura, P T Rigney and V Stannett,
    J. Polym. Sci. (C), 1969, 28, 243.

45. P P Roussis, Polymer, 1981, 22, 768.

46. P P Roussis, Polymer, 1981, 22, 1058.

47. H G Burghoff and W Pusch, J. Appl. Polym. Sci., 1976, 20,
    798.

48. H G Burghoff and W Pusch, J. Appl. Polym. Sci., 1979, 23, 473.

49. H G Burghoff and W Pusch, Polym. Eng. Sci., 1980, 20, 305.

50. C Toprak, J N Agar and M Falk, J. Chem. Soc., Faraday Trans I,
    1979, 75, 803.

51. J R Scherer and G F Bailey, J. Membrane Sci., 1983, 13, 43.

52. J A Sauer and L S A Smith, Trans. I. Mar. E.(C) 97 Conf. 2
    1985, 95.

53. L S A Smith, Polymer, 1981, 22, 822.

54. S G Kimura and G E Walmet, Separation Sci. and Tech., 1980, 15, 1115.

55. European Patent Office Publ. No. 0159783, 1985.

56. F A Long and L J Thompson, J. Polym. Sci., 1954, 14, 321.

57. J A Myer, C Rogers, V Stannett and M Szware, Tappi, 1957, 40, 142.

58. V L Simril and A Hershberger, Modern Plastics, July, 1950,95.

59. N T Norley, J. Apply. Chem., 1963, 13, 107.

60. F L Pilar, J. Polym. Sci., 1960, 45, 205.

61. Y Ito, Chemistry High Polym, (Japan) 1961, 18, 158.

62. W B Kunz and R T K Cornwell, Tappi, 1962, 45, 583.

63. Y Kamiya and F Takahashi, J. Appl. Polym. Sci. 1977, 21,1945.

64. Y Kaminya and F Takahashi, J. Appl. Polym. Sci., 1979, 23, 627.

65. G T Paulson, A B Clinch and F P McCandless, J. Membrane Sci., 1983, 14, 129.

66. D G Pye, H H Hoehn and M Panar, J. Appl. Polym. Sci., 1976, 20, 287.

67. H G Spencer and I M Ibrahim, J. Appl. Poly. Sci., 1978, 22, 3607.

68. R T Chern, W J Koros, E S Sanders and R Yui, J. Membrane Sci. 1983, 15, 157.

69. R M Barrer and J A Barrie, J. Polym. Sci. 1957, 23, 331.

70. J A Barrie, P S Sagoo and P Johncock, J. Membrane Sci. 1984, 18, 197.

71. J A Barrie, P S Sagoo and P J Johncock, Polymer, 1985, 26, 1167.

72. W J Koros and D R Paul, J. Polym. Sci. Polym. Phys. Ed., 1978, 16, 1947.

73. A G Wonders and D R Paul, J. Membrane Sci. 1979, 5, 63.

74. J M Fletcher, H B Hopfenberg and W J Koros, Polym. Engr. and Sci., 1981, 21, 925.

75. R T Chern, W J Koros, E S Sanders, S H Chen, and H B Hopfenberg. In "Industrial Gas Separations", ACS Symp. Ser. 223, 1983, 47.

76. M Kawaguchi, T Tanguchi, K Tochigi and A Takizawa, J. Appl. Polym. Sci., 1975, 13, 493.

77. W H Hubbel, H Brandt and Z A Munir, J. Polym. Sci., Polym. Phys. Ed., 1875, 13, 493.

78. J A Barrie, P S Sagoo and P Johncock, J. Appl. Polym. Sci., in press.

79. B P Tikhomirov, H B Hopfenberg, V Stannett and J L Williams, Die Makromol. Chemie, 1968, 113, 177.

80. N Minoura and T Nakagawa, J. Appl. Polym. Sci., 1979, 23, 2729.

81. V B Singh and D J Walsh, J. Macromol Sci. Phys., 1986, B25, 65.

82. S M Allen, M Fujii, V Stannett, H B Hopfenberg and J L Williams, J. Membrane Sci. 1977, 2, 153.

83. V Stannett, G R Ranade and W J Koros, J. Membrane Sci., 1982, 10, 219.

84. V Stannett, M Haider, W J Koros and H B Hopfenberg, Polym. Eng. and Sci., 1980, 20, 300.

85. R M Ikeda and F P Gay, J. Appl. Polym. Sci., 1973, 17, 3821.

86. A G Day, Trans. Faraday Soc., 1963, 59, 1218.

87. J A Barrie and K Munday, J. Membrane Sci., 1983, 13, 175.

88. G J van Amerongen, Rubber Chem. and Tech. 1964, 37, 1067.

89. J A Barrie and J Taylor, unpublished results.

90. V Stannett, In "Diffusion in Polymers" (J Crank and G S Park eds) Academic Press, London, 1968, Chapter 2.

91. K Toi, G Morel, D R Paul. J. Appl. Polym. Sci. 1982, 27, 2997.

92. R T Chern, W J Koros, H B Hopfenberg and V T Stannett, In "Material Science of Synthetic Membranes" ACS Symp. Ser.269, 1985, 25.

93. Y Maeda and D R Paul, Polymer, 1985, 26, 2055.

94. T Asada and S Onogi, J. Colloid Sci, 1963, 18, 784.

95. K Toi, Y Maeda, T Tokuda, J. Membrane Science, 1983, 13, 15.

96. K Toi, Polym. Eng. and Sci., 1980, 20, 30.

97. R T Chern, W J Koros, E S Sanders and R Yui, J. Memb, Sci., 1983, 15, 157.

98. J A Barrie, H J Rudd, P S Sagoo and P Johncock, Br. Polym. J., in press.

99. P Johncock and G F Tudgey, Br. Polym. J., 1983, 15, 14.

Polymers for Gas Separation Membranes

By I. K. Ogden*, R. E. Richards, and A. A. Rizvi

BRITISH PETROLEUM COMPANY PLC, RESEARCH CENTRE, CHERTSEY ROAD, SUNBURY, MIDDLESEX, TW16 7LN, UK

1. INTRODUCTION

Polymeric membranes are being increasingly used to effect separations of gas streams in a variety of applications. Examples of such applications would include the generation of medium purity (ca 97%) nitrogen from compressed air, the recovery of hydrogen from refinery purge gases and the removal of carbon dioxide from produced natural gas streams. In many cases the polymers employed for the fabrication of the membrane are glassy amorphous materials characterised by high glass transition temperatures, good mechanical strength and an acceptable combination of gas permeability and selectivity properties.

The permeability of a polymer to gases depends upon both the physical properties of the polymer and the gases concerned. For a particular gas the nature of the polymer and its interaction with the gas will clearly determine the transport behaviour. Factors which relate to the molecular structure of the polymer, such as polarity, hydrogen bonding, cohesive energy density, chain flexibility, steric hinderance, and crystallinity will (amongst others) all have an influence on the transport process(1). It is evident that in correlating gas solubility and diffusivity with polymer structural properties it is difficult to isolate these many inter-relating factors. The selectivity of a polymer to a particular gas mixture is a still more complicated issue, because gases behave competitively in glassy polymer systems.

In developing a clearer understanding of factors affecting glassy polymer gas selectivity, careful studies of pure gas sorption and permeation are required. Analysis of comparative gas sorption and transport behaviour should result in the development of a clearer appreciation of the inter-relationship between polymer structural factors and gas separation behaviour. Such studies can usefully be carried out within the theoretical framework provided by, for example, the dual-mode sorption model. In this paper, the sorption, permeation and selectivity behaviour of three commercially available amorphous polymers is reported and compared.

2. BACKGROUND: DUAL-MODE SORPTION MODEL

The sorption and permeation of gases and vapours in glassy polymers has been shown to be well described over widely varying conditions by the dual-mode sorption theory(2-5) ie,

$$C = C_D + C_H \qquad [1]$$

$$C_D = k_D P_\bullet \qquad [2]$$

$$C_H = \frac{C'_H b P_\bullet}{1 + b P_\bullet} = \frac{K C_D}{1 + \beta C_D} \qquad [3]$$

$$K = \frac{C'_H b}{k_D} \qquad \beta = \frac{b}{k_D}$$

114

where

C = sorbate concentration/cm^3 (stp) cm^{-3} (polymer)
C_D = dissolution sorption (Henry's Law)
C_H = microvoids sorption (Langmuir isotherm)
C'_H = microvoid capacity constant/cm^3(stp)cm^{-3}(polymer)
k_D = Henry's Law distribution constant/cm^3(stp)cm^{-3}(polymer) bar^{-1}
$P_\bullet$ = sorbate equilibrium pressure/bar
b = hole affinity constant/bar^{-1}

The permeability of a gas P can be shown(6) to be given by the expression:

$$P = k_D D \left[1 + \frac{FK}{1 + bP_2} \right] \qquad [4]$$

where $D = D_D$, $F = D_H/D_D$, P_2 is the upstream gas pressure,

and D_D, D_H are diffusion coefficients associated with the Henry's law and Langmuir mode sorbed populations, respectively.

Equations [3] and [4] successfully describe the observed characteristics of gas-glassy polymer systems, viz non-linear solubility behaviour and pressure dependent permeabilities will arise. The extent of the deviations will depend upon the extent of sorption onto Langmuirian sites relative to dissolution into the matrix of the polymer, and the relative mobilities of these two types of species.

For rubbery polymers, where gas diffusion is Fickian, an ideal separation factor α^* for gases A, B is given by:

$$\alpha^* = \frac{P_A}{P_B} = \frac{k_{DA}}{k_{DB}} \cdot \frac{D_A}{D_B} \qquad [5]$$

$$= \widehat{S}(A/B) \cdot \widehat{D}(A/B)$$

where $S(A/B)$, $D(A/B)$ are the solubility selectivity and mobility (diffusivity) selectivity, respectively for gas pair A, B.

For glassy polymers relationship [5] may be used, to a first approximation, to obtain the separation factor α (2):

$$\alpha = S(A/B) \cdot D(A/B) \qquad [6]$$

In this case S and D are solubility and mobility selectivities of a glassy polymer which are expected to be pressure dependent (cf equations [3] and [4]). It is evident from equation [6] that the ability of glassy polymers to separate gases is dependent upon contributions from both solubility and mobility selectivities. In the case of CO_2/CH_4 separation, Koros (2) has argued that mobility selectivity plays the dominant role. It should be noted, however, that equation [6] is only valid at high pressures.

3. MATERIALS

Kapton film was obtained from Dupont UK Ltd, Hemel Hempstead (film type 30H) and was used as received. Ultem (General Electric) and polyethersulphone (PES, obtained from Aldrich Chemicals) films were prepared by casting solutions in chloroform and tetrahydrofuran, respectively, onto clean mercury surfaces. These films were pretreated in a vacuum oven at temperatures just below the glass temperature, in order to remove residual solvent and to anneal the films. Properties of the films used are given in Table 1.

All gases were supplied by BOC (CP grade) and were used without further purification.

TABLE 1

PHYSICAL PROPERTIES OF FILMS USED IN SORPTION WORK

Polymer	Average film thickness/μm	Density* /g.cm^{-3}	Film weight /mg
Kapton	7.5	1.435	46.2
Polyethersulphone	70	1.229	419.3
Ultem	51 and 20	1.281	357.1 and 120.4

*at room temperature

4. EXPERIMENTAL

(a) Sorption Isotherms

Sorption isotherms were measured gravimetrically using a specially constructed apparatus employing a Sartorius equal-arm balance. The details of this apparatus, possible sources of error, and its mode of operation will be presented elsewhere (7). In general, isotherms were reproducible to within ca 2%. Films were outgassed by evacuating the system for a period > 48 hours at 60°C. Essentially, the mass change for sorption equilibrium relative to the sample mass under vacuum was recorded as a function of the gas pressure P_e at constant temperature. Sorption equilibrium was established in < 4 hours for CO_2 in Ultem and PES and in less than 8 hours for the Kapton film. The original weight uptake data were corrected for buoyancy and electronic drift (if applicable) and converted to units of concentration (cm^3(stp) cm^{-3} (polymer)) and sorbate molecules per repeat unit (m/ru).

(b) Permeation Measurements

Pure gas permeation measurements were carried out in a constant volume apparatus. The permeation cell was of all metal construction, with the membrane being supported by a filter paper resting in turn upon a fine steel mesh screen. Films were degassed by evacuating to <1x10^{-4} Torr at least overnight. Pressure was maintained essentially constant on the upstream face of the membrane during the course of a permeation experiment, whilst the downstream pressure was < 1% the upstream pressure. The permeation cell could be thermostated to the required temperature. Steady-state permeabilities were measured by following the rate of pressure increase, using a sensitive Baratron pressure transducer, in a calibrated downstream buffer volume. Permeabilities were reproducible to within ca 5%.

5. RESULTS

   (a) Sorption

   Sorption isotherms are presented in Figures 1 - 4 for the following systems:

   (i)   Polyethersulphone (PES) - CO_2 at 40.0, 50.2 and 60.0°C (Fig 1)
   (ii) PES-CH_4 at 40.0, 50.0 and 60.0°C (Fig 2)
   (iii) Ultem (polyetherimide) - CO_2 at 35.0 and 60.0°C (Fig 3)
   (iv) Kapton (polyimide) - CO_2 at 60.0°C (Fig 3)
   (v)   Ultem - CH_4 at 35.0 and 60.0°C (Fig 4)

   Sorbate concentrations were found to be in good agreement with literature data for PES-CO_2 (8) and Kapton-CO_2 (3). Calculation of Kapton-CO_2 sorption data from the dual mode sorption parameters reported in reference (3) indicated agreement to be within 14%, which is acceptable given the large differences in experimental techniques and the differing film samples, and is also within the quoted errors of the dual-mode parameters. Methane sorption was not studied on Kapton, but instead an isotherm was calculated on the basis of literature data (3) according to the dual-mode sorption equation [1] - see section 6.a.

   As expected all the isotherms display distinct curvature, typical of dual-mode sorption. PES sorbs less CO_2 and CH_4 than Ultem at 60°C for a given pressure. At this temperature CO_2 sorption levels within Ultem and Kapton are approximately equal. It is interesting, however, to calculate the CO_2 uptake of these polymers in terms of molecules of gas sorbed per monomer repeat unit (m/ru), on the basis of the repeat unit structures illustrated in Figure 5. The isotherms have been replotted in Figure 6. The ranking of polymer CO_2 sorption at 60°C is now Ultem > PES ~ Kapton.

   (b) Permeability and Selectivity of Ultem Films

   Carbon dioxide and methane permeabilities were measured for Ultem at 30°C and 60°C. In Ultem CO_2 permeabilities showed pressure dependence, decreasing slightly with pressure in the range 0.5 to 6 Bar. Zero pressure permeabilities were extrapolated from the data at 30°C and values of 1.53×10^{-10}, 1.59×10^{-10} and 1.50×10^{-10} cm³ (stp)cm/(cm².s.cmHg) obtained for three films 53, 39 and 21 µm in thickness, respectively. In these films methane permeabilities showed little pressure dependence over a similar range of pressures. Permeability values extrapolated to zero pressure for the same three films were 3.5×10^{-12}, 4.0×10^{-12} and 3.3×10^{-12} cm³ (stp) cm/(cm².s.cmHg) at 30°C. The wider variation in the methane data is due to the greater significance of the leak rate correction on the downstream side, resulting from lower fluxes.

   From these data ideal zero pressure selectivities for the CO_2/CH_4 gas pair of ca 44, 40 and 45 were obtained.

   Permeation data for the 39 µm thick Ultem film were also obtained at 60°C; zero pressure permeabilities obtained for CO_2 and CH_4 were 2.5×10^{-10} and 1.3×10^{-11} cm³ (stp) cm/(cm².s.cmHg) respectively, giving a CO_2/CH_4 ideal selectivity of ca.19 for this gas pair.

   The permeability and selectivity data are summarised in Table 2.

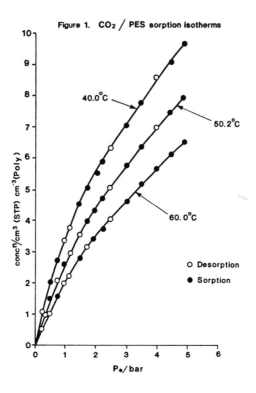

Figure 1. CO_2 / PES sorption isotherms

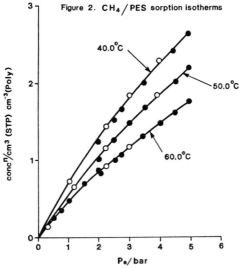

Figure 2. CH_4 / PES sorption isotherms

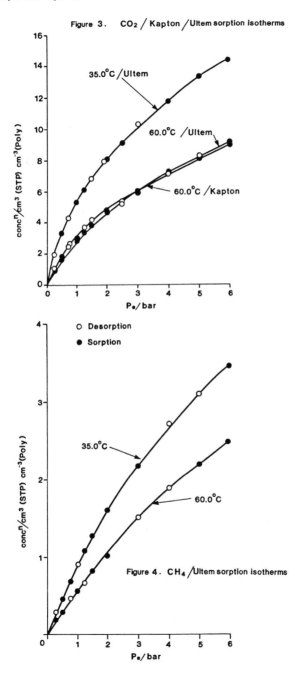

Figure 3. CO_2 / Kapton /Ultem sorption isotherms

35.0°C /Ultem

60.0°C /Ultem

60.0°C /Kapton

$conc^n/cm^3$ (STP) cm^{-3}(Poly)

P_e/bar

O Desorption
● Sorption

35.0°C

60.0°C

Figure 4. CH_4 /Ultem sorption isotherms

$conc^n/cm^3$ (STP) cm^{-3}(Poly)

P_e/bar

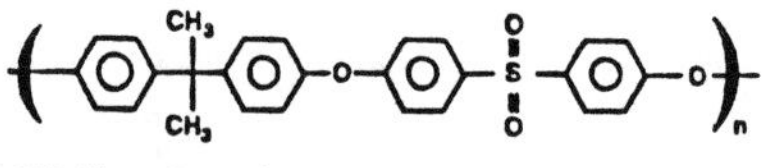

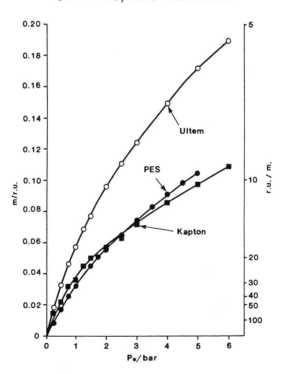

PES (Union Carbide)

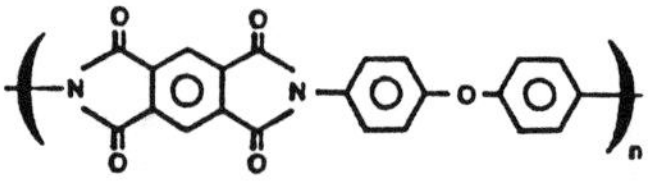

Kapton (Dupont)

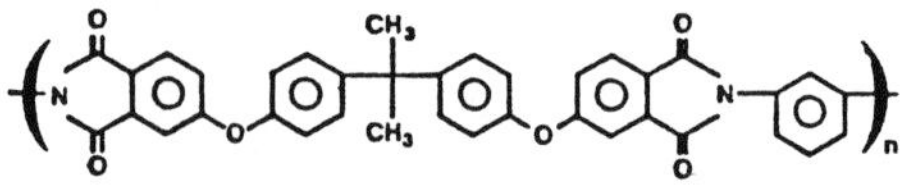

Ultem (GEC)

Figure 5. Polymer repeat units

Figure 6 . CO_2 / 60.0°C   sorption isotherms

TABLE 2

PERMEATION DATA FOR ULTEM FILMS

Polymer	Permeability at 30°C /cm³(stp)cm/(cm².s.cmHg)		$\propto(CO_2/CH_4)$ at 30°C
	CO_2	CH_4	
Ultem 1	1.5×10^{-10}	3.5×10^{-12}	44
Ultem 2	1.6×10^{-10}	4.0×10^{-12}	40
Ultem 3	1.5×10^{-10}	3.3×10^{-12}	45

6. DISCUSSION

(a) CO_2/CH_4 Solubility Selectivity

Figure 7 shows the solubility selectivity of Ultem and PES for CO_2 relative to methane ($S(CO_2/CH_4) \equiv C(CO_2)/C(CH_4)$ cf equation [6]). It is evident that the solubility selectivity decreases with increasing P_e. Furthermore, it is striking that PES and Ultem possess a remarkably similar selectivity at 60°C for $P_e > 2.5$ bar, eg, $S(CO_2/CH_4) = 3.77$ and 3.74 for Ultem and PES at $P_e = 5$ bar. In terms of solubility selectivity, $S(CO_2/CH_4)$, at 60°C, PES and Ultem would appear to have the same effectiveness, although the sorbate uptake is higher for Ultem than for PES. In addition, the solubility selectivity of Ultem decreases with increased sorbent temperature. This implies that increasing the temperature causes the CO_2 solubility to decrease to a greater extent than that of CH_4, which in turn indicates that the heat of solution, ΔH_s, for CO_2 is more exothermic than that of CH_4.

Carbon dioxide and methane sorption data in Kapton at 60°C were calculated on the basis of the dual-mode parameters provided in reference (3). From the resulting isotherms $S(CO_2/CH_4)$ at 60°C was calculated. These data are also shown in Figure 7; it can be seen that Kapton has the highest solubility selectivity of the three polymers studied, and is higher at 60°C than that of Ultem at 35°C. The $S(CO_2/CH_4)$ for Kapton decreases with increasing pressure but remains well above the value attained by Ultem and PES at all pressures within the range of measurement. Kapton would thus appear to be more selective on a sorption basis than Ultem in particular, in spite of the fact that sorbate uptakes are very similar. Kapton is not expected to decline to $S(CO_2/CH_4)$ values attained by PES and Ultem at higher pressures, because of the molecular structural differences between Kapton and the other two polymers.

The observed permeation selectivity of Ultem for CO_2/CH_4 has an average value of 43 at zero pressure and 30°C (Table 2). By application of relationship [6] it is clear that the sorption selectivity provides a significant contribution to this overall selectivity.

Permeation data for PES have been reported in the literature (8) at 35°C, and analysed according to the dual-mode theory. From the parameters quoted an estimate was made of the zero pressure permeability of CO_2 and CH_4 (Table 3) at this temperature, and consequently an ideal zero pressure selectivity for CO_2/CH_4 of 32 calculated. Koros (2) reports a selectivity of 28 at 35°C and 20 bar. From Figure 7 it is evident that the solubility selectivity plays a significant, but not dominant role in determining the overall CO_2/CH_4 permselectivity of PES and Ultem.

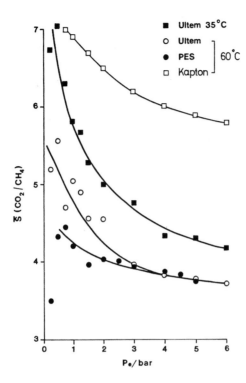

Figure 7. CO_2/CH_4 solubility selectivity

A similar technique was used to determine the zero pressure permeabilities of Kapton at 60°C (Table 3). The selectivity estimated on this basis is 44 which is somewhat lower than, but not inconsistent with the value of 63.6 at 35°C and 20 atm reported elsewhere (9). Once again, the mobility selectivity contribution to the overall selectivity should exceed $S(CO_2/CH_4)$.

TABLE 3

PERMEABILITIES OF POLYETHERSULPHONE AND KAPTON FILMS
CALCULATED AT ZERO PRESSURE FROM DUAL MODE PARAMETERS

Polymer	Permeability $cm^3(stp)cm/(cm^2.s.cmHg)$		CO_2/CH_4 Selectivity
	CO_2	CH_4	
Polyethersulphone[1] (35°C)	7.4×10^{-10}	2.3×10^{-11}	32
Kapton (60°C)[2]	4.9×10^{-11}	1.1×10^{-12}	44
Kapton (35°C)[3]	2.0×10^{-11}	3.2×10^{-13}	64

1. Data from reference (8)
2. Data from reference (3)
3. Values at 20 atm, data from reference (9)

(b) Relative Sorbate Uptake

The variation in the CO_2 sorbate concentration with P_e, at 60°C, for Ultem relative to PES and for Ultem relative to Kapton is presented in Figures 8 and 9. Interestingly, linear relationships are obtained for the P_e range investigated (a similar correlation is produced for sorbate concentration based on m/ru calculations, although with a different gradient). Furthermore, the extrapolation to zero P_e does not intercept the origin. Presumably, a detailed characterisation of the low P_e (<0.25 bar) sorption region would indicate a deviation from the linear correlation. The linearity of the relationships presented in Figures 8 and 9 is remarkable, considering (i) the curvature of the isotherms, and therefore (ii) both the pressure dependence of ratios of polymer solubility, and the interpolymer selectivity as defined below.

(c) Interpolymer Selectivity

The interpolymer selectivities, Spoly, for both CO_2 and CH_4 sorption (60°C) are established from the sorbate uptake (C_j) for each sorbate-polymer system, as a function of $P_{e,j}$. For example, the interpolymer selectivity of polymers E and F for sorbate A is given by:

$$\overline{S}poly = \frac{E}{F} \overline{S}(A) = \frac{C_j{}^E}{C_j{}^F} \quad \text{at } P_{e,j} \qquad [7]$$

Interpolymer selectivities are presented in Figure 10 (cm^3 (STP) cm^{-3} (polymer)) and Figure 11 (m/ru).

The important factors to emerge from Spoly, in the P_e range investigated were:

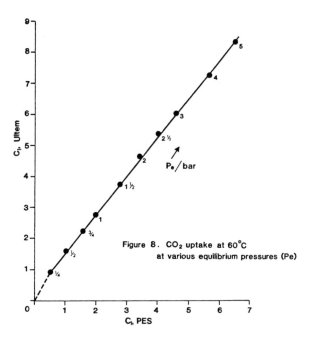

Figure 8. CO_2 uptake at $60°C$ at various equilibrium pressures (Pe)

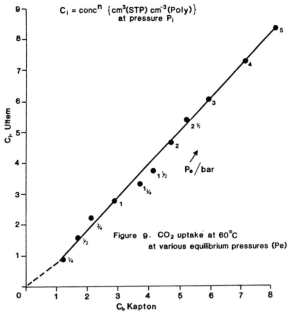

$C_i = conc^n \{cm^3(STP)\, cm^{-3}(Poly)\}$ at pressure P_i

Figure 9. CO_2 uptake at $60°C$ at various equilibrium pressures (Pe)

(NB C indicates cm^3(STP)cm^{-3}(polymer)
     m/ru indicates sorbate molecules per repeat unit)

(i) Spoly is P$_•$ dependent for both CO$_2$ and CH$_4$.

        Ultem   Ultem
(ii) S Kapton (CO$_2$) and S PES (CH$_4$) increase at low P$_•$ to an
      approximately constant value for P$_•$ $\geq$ 2 bar and P$_•$ $\geq$ 3 bar
      respectively.

        Ultem   Kapton
(iii) S PES (CO$_2$) and S PES (CO$_2$) decrease at low P$_•$ to an
       approximately constant value for P$_•$ $\geq$ 3 bar.

and, on a sorbate concentration basis (cm^3(STP)cm^{-3}(polymer)):

        Ultem
(iv) at P$_•$ $\geq$ 2 bar, S Kapton (C, CO$_2$) ~1. At high pressures
      therefore, the Ultem and Kapton matrices have an equal affinity
      for CO$_2$ on a concentration basis.

      Ultem   Kapton   Ultem
(v) at P$_•$ > 3 bar, S PES (C, CO$_2$) ~ S PES (C, CO$_2$)~ S PES (C,CH4) ~
     1.25

and, on a molecular basis, considering molecules sorbed per repeat unit:

        Ultem
(vi) at P$_•$ > 2 bar, S Kapton (m/ru, CO$_2$) $\simeq$ 1.75, therefore Ultem has a
      greater sorption affinity per polymer repeat unit for CO$_2$
      relative to Kapton.

       Ultem   Ultem
(vii) at P$_•$ > 3 bar, S PES (m/ru, CO$_2$) ~ S PES (m/ru, CH$_4$)~ 1.65
      Kapton
       and S PES (m/ru, CO$_2$) ~ 0.95, therefore Ultem for CO$_2$ and CH$_4$ has
       a greater, and Kapton for CO$_2$ has an approximately equivalent
       sorption affinity per polymer repeat unit relative to PES.

       Ultem   Kapton
(viii) S PES (m/ru, CO$_2$) > S PES (m/ru, CO$_2$) for all P$_•$ studied.

(ix) the number of sorbate molecules per polymer repeat unit is very
      small, cf Figure 6. For example, for PES-CO$_2$ at P$_•$ = 5 bar, the
      sorbate uptake is approximately 0.1 m/ru, or alternatively, 1 CO$_2$
      sorbed per every 10 PES repeat units.

(d) Sorption Sites and Accessibility

    In a recent review of gas solubility behaviour (2), carbonyl
(-CO-) and sulphonyl (-SO$_2$-) groups were considered as strong sorption sites
with an approximately equivalent solubility selectivity for CO$_2$/CH$_4$. By
comparison of the Ultem, Kapton, and PES polymer repeat units, Figure 5, Ultem
and Kapton are identical in the number of strong sorption sites per repeat unit
(4x -CO-), whereas in contrast PES possesses only a single strong sorption site
per repeat unit (1x -SO$_2$-). Therefore, the relative CO$_2$ sorption uptake, on
consideration of the number of strong sorption sites per polymer repeat unit,
should be 4:1 for (Ultem/Kapton) : PES at a given P$_•$ and temperature.

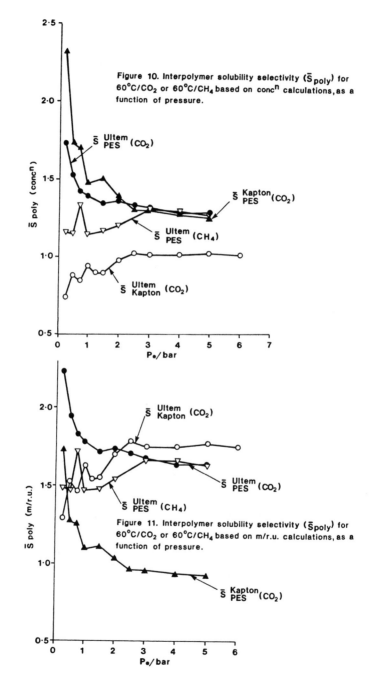

Figure 10. Interpolymer solubility selectivity ($\bar{S}_{poly}$) for $60°C/CO_2$ or $60°C/CH_4$ based on $conc^n$ calculations, as a function of pressure.

Figure 11. Interpolymer solubility selectivity ($\bar{S}_{poly}$) for $60°C/CO_2$ or $60°C/CH_4$ based on m/r.u. calculations, as a function of pressure.

However, the interpolymer selectivities calculated from the m/ru sorption data (Figure 11) provided values of 2.2 to 1.6:1 for Ultem : PES and 1.7 to 0.9:1 for Kapton:PES, in poor agreement with the anticipated values of 4:1. Therefore, a statistical interpretation is oversimplistic.

The possibility exists that each -CO- group of an imide ring does not constitute a single strong sorption site. The close spacing resulting from the five membered imide ring may prevent simultaneous sorption at each carbonyl group. Therefore, the whole imide ring may provide a single strong sorption site and so both Ultem and Kapton would contain twice the number of sites relative to PES. The observed m/ru interpolymer selectivity for CO_2 is in moderate agreement with this conjecture. However, note that a CO_2 may possibly sorb both sides of the imide ring, simultaneously.

Of fundamental importance is the accessibility of the strong sorption sites, since in a dense polymer film strong sorption sites also act as centres for interchain attraction. Therefore a sorbate would presumably have to disrupt the interchain packing to sorb at these specific sites.

The degree of order in polymer chain packing may be considered to be dependent upon

(i) flexibility/rigidity of polymer chains.

(ii) density of interchain attraction centres, and the magnitude of the interaction.

Both Ultem and PES contain three flexible linkages per repeat unit (2x -O-, 1x -C(CH3)2-). In contrast, Kapton possesses only a single flexible linkage per repeat unit, Figure 5: (1x -O-). Furthermore, the density of the interchain attraction centres decreases in the order Kapton > Ultem > PES, since Kapton is a smaller polymer repeat unit in relation to Ultem. The presence of molecular ordering in Kapton (parallel chain alignment) has been reported from X-ray diffraction studies (10). Overall, the expected randomness of chain packing is PES > Ultem > Kapton and, similarly for polymer film density PES < Ultem < Kapton (as measured).

The Pe dependence, or sorbate uptake dependence, of S(CO2/CH4) and Spoly, may be associated with the accessibility of strong sorption sites. The interaction energy of CO_2 relative to CH_4 with a strong (specific) sorption site may be considered to be greater, and for CH_4 a greater influence from non-specific interactions would consequently be envisaged. The excess free volume of unrelaxed glassy polymers may tentatively be related to the accessibility of the strong sorption sites. At very low Pe and sorbate concentrations (Figure 11), the CO_2 interpolymer selectivities (to a first approximation) tend to those values expected statistically (as above).

ie   S Kapton (m/ru, CO_2) tends to unity, and, S PES (m/ru,CO_2) ~

    Ultem
    S PES (m/ru, CO_2) tends to 2-4.

However, as the sorption level is increased, the accessible strong sorption sites are saturated, and sorption occurs increasingly at less energetically favourable sites, through non-specific Henry's Law type dissolution.

In summary:

(i) S PES (m/ru, CO_2), and S PES (m/ru, CO_2) decrease with increased
       Ultem   Kapton
                      Ultem
     P_e, and S Kapton (m/ru, CO_2) increases with increased P_e reflecting
the "ordered chain packing" related accessibility of strong sorption
sites.

(ii) S PES (m/ru, CH_4) increases slightly, at increased P_e, reflecting the
               Ultem
degree of interchain attraction and the influence of the non-specific
interaction of CH_4 with a polymer repeat unit.

(iii) The interpolymer sorption selectivity and the CO_2/CH_4 sorption
selectivity reach constant values as the sorbate uptake increases.
This is because of the saturation of selective specific sorption
sites, and the increased role of the less selective/non-specific
sorption sites of a repeat unit, eg ether link.

(e) Correlation with Physical Parameters

    Solubility paramters (δ), glass transition temperatures (T_g) and
densities (ρ) of the polymers are given in Table 4 together with the lower
temperature (30/35°C) separation selectivity and permeability data. It is
evident that selectivities decrease with decreasing T_g and ρ, whilst
permeabilities increase with decreasing selectivity for the polymers studied.

TABLE 4

CORRELATION OF SOME PHYSICAL PROPERTIES WITH PERMSELECTIVITY

Polymer	Solubility parameter $(Jcm^{-3})^{1/2}$	Glass Transition /°C	Density /gcm^{-3}	α	CO_2 Permeability $cm^3(stp)/cm.s.cmHg$
Kapton	24.6	400–500[3]	1.435	64	2.0×10^{-11}
Ultem	21.4[2]	215[2]	1.281	44	1.5×10^{-10}
PES	21.3[2]	180–185[4]	1.229	32	7.4×10^{-10}

1. Data from ref (9)   3. Data from ref (3)
2. Data from ref (11)   4. Data from ref (8)

7. CONCLUSIONS

    (i) Of the three polymers studied Ultem is the most effective CO_2 sorbent
on the basis of molecules of CO_2 sorbed per polymer repeat unit. This is true
over the whole range of the measurements (0-6 bar) investigated. However, it

is interesting to note that overall levels of sorption are low ie only one molecule of CO_2 is sorbed every five Ultem monomer units at 6 bar equilibrium pressure.

(ii) A remarkable similarity in CO_2/CH_4 solubility selectivity is observed in this study for Ultem and PES (for $P_{\bullet} > 2.5$ bar). Consequently it is proposed that for these systems mobility selectivity plays a significant role in determining overall selectivity.

(iii) Interpolymer solubility selectivity results indicate that imide rings in Kapton and Ultem, and $-SO_2-$ groups in polyethersulphone act as strong specific sorption sites.

(iv) Kapton has the highest CO_2/CH_4 selectivity of the three polymers studied, and also the lowest permeability. Low permeabilities, combined with lack of organic solvent solubility may render development of Kapton membranes difficult.

8. ACKNOWLEDGEMENT

The authors would like to thank the British Petroleum Company for permission to publish this paper, and Ms P.J. Hart for expert technical assistance. Valuable discussions with Dr W.D. Webb are also gratefully acknowledged.

8. REFERENCES

(1) Crank, J., Park, G.S., "Diffusion in Polymers", Academic Press London, (1968).

(2) Koros, W.J., J.Poly.Sci.Poly.Phys. Ed., 23, 1611 (1985).

(3) Chern, R.T., Koros, W.J., Yui, E., Hopfenberg, H.B., and Stannett, V.T., J.Poly.Sci.Poly.Phys. Ed., 22, 1001 (1984).

(4) Vieth, W.R., Howell, J.M., Hsieh, J.H., J.Mem.Sci., 1, 177 (1976).

(5) Sangani, A.S., J.Poly.Sci. Poly.Phys. Ed., 24, 568 (1986).

(6) Koros, W.J., Chan, A.H., Paul, D.R., J.Mem.Sci., 2, 165, (1977).

(7) Richards, R.E., Rizvi, A.A., Webb, W.D., to be published.

(8) Erb, A.J., Paul, D.R., J.Memb.Sci., 8, 11 (1981).

(9) Chern, R.T., Koros, W.J., Hopfenberg, H.B., Stannett, V.T., in "Materials Science of Synthetic Polymer Membranes", Ed Lloyd, D.R., ACS Symp. Series, 269, p.25, (1985).

(10) Takahashi, N., Yoon, D.Y., Parrish, W., Macromolecules, 17, 2583, (1984).

(11) Kambour, R.P., Polymer Commun., 24, 292 (1983).

Production of Porous Hollow Polysulphone Fibres for Gas Separation

By G. C. East, J. E. McIntyre*, V. Rogers, and S. C. Senn

DEPARTMENT OF TEXTILE INDUSTRIES, THE UNIVERSITY OF LEEDS, LEEDS LS2 9JT, UK

Introduction

The basic requirements for membranes suitable for gas separation by permeation processes are high permeability to and high selectivity in favour of the desired component of the mixture. These two requirements are generally in conflict. Choice of materials for flat and spiral sheet membranes is particularly limited by this conflict between permeability and selectivity. Hollow fibre membranes are much less limited because their greater surface area per unit volume permits the practical use of structures of lower permeability in area terms while maintaining a high permeation flow in volume terms.[1] Moreover, hollow fibre membranes can remain intact at high pressure differentials across the membrane without additional support, which is often necessary for the sheet membranes. It is necessary, however, to ensure that the internal diameter of the fibres is sufficient to prevent a substantial pressure drop along the bore of the fibre (i.e. on the permeate side).[2]

Hollow fibre permeators have previously been used in commercial reverse osmosis systems, for example in desalination processes, and also in ultra-filtration, in dialysis, and in ion exchange systems. Hollow fibres used for separations involving liquids are often too porous to be effective gas separators. Hollow fibres designed for gas separation have a porous internal wall structure together with a thin, relatively non-porous, skin but even these may possess a small but significant pore structure that extends through the skin to the surface and prevents the membrane from reaching its full potential in terms of selectivity.

Application of a very thin coating of a highly permeable polymer that blocks the pores has proved a very useful technique for improving selectivity without an excessive fall in permeation rate. The technique has been applied particularly to cellulose acetate,[3,4] polysulphone[3,5] and acrylonitrile/styrene copolymer[3,6] hollow porous fibres. Typically the coating polymer is elastomeric, usually a silicone. After the coating has been applied the substrate, not the coating, becomes the effective separating material,[7] and a model for such composite membranes predicts that for intermediate surface porosities in the fibre before coating, the selectivity after coating should be an order of magnitude better than before coating, and not much less than that for uncoated fibres of very low surface porosity. Hollow porous fibres based on a polysulphone fibre coated with a silicone polymer are reported to be the basis of the original form of the Monsanto PRISM gas separation system.[8]

Although the production of hollow porous polysulphone blend fibres for liquid permeation processes has been described in the general literature,[9,10] most of the information about the production of fibres for gas separation is to be found in patents.[3,11] This paper describes an investigation of a process for making such fibres and of the properties of the resulting product.

Materials

The majority of the work described was carried out using P3500 poly-sulphone from Union Carbide. Victrex 200P polyethersulphone from I.C.I. was also examined. These polymers have the following chemical repeating units:-

P3500

Victrex 200P

Procedure for making hollow porous fibres

Figure 1 shows schematically the spinning arrangement used for making membrane fibres. A pre-formed solution of a polymer in an appropriate solvent was fed under nitrogen pressure from a heatable reservoir at controlled temperature to a gear metering pump. The rate of rotation of the gears controlled the feed of polymer solution through a filter to the spinneret. The spinneret, of a single hole tube-in-orifice type (Plate 1), was one of a number of spinnerets with different orifice diameters (O.D.), tube outer diameters (T.D.) and inner diameters (I.D.) ranging from 570/255/150 μm to 820/430/340 μm. The internal coagulant, which was deionised water for all the work described in this paper, was injected down the tube at a controlled rate from either a syringe pump (batchwise) or more usually a peristaltic pump (continuously). For any single spinneret the linear extrusion rate (L.E.R.)

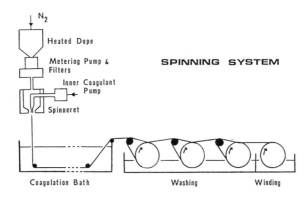

Figure 1.   Spinning arrangement for making membrane fibres.

of polymer is proportional to the polymer volume extrusion rate (P.E.R.) and the linear extrusion rate of water is proportional to the water volume injection rate (W.I.R.), but smaller holes lead to higher linear rates relative to the volume rates.

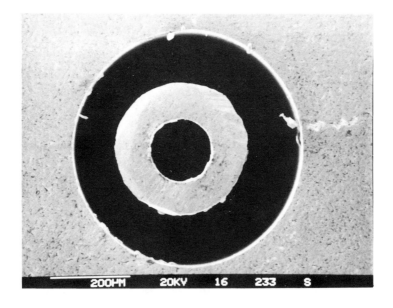

Plate 1.   Face view of tube-in-orifice spinneret.

The spinneret was arranged to extrude vertically downwards and was either immersed in the coagulant so that coagulation began as the polymer stream left the orifice or was situated above the coagulant so that the filament passed through an air gap before entering the coagulant.   Operation with an air gap facilitates operation with a temperature difference between the extrudate and the coagulant.

The coagulant was water in all the work described here.   The filament passed vertically through the coagulant to a guide, then horizontally to a second guide and at an angle upwards to the first of a series of three rotating Neoprene-covered drums round each of which it passed several times and on which it was washed by a spray of water.   The filament was then wound up on a drum winder as a band of non-overlapping filaments which were cut from the drum to give hollow fibres of length 70 cm.

It was found to be essential to use low friction guides to avoid damaging or distorting the fibres.   Use of a cone-type winder, which led to overlapping filaments on the package, was unsatisfactory because of crushing at the points of overlap and resulting variations in filament diameter.

After cutting from the drum, the filaments were suspended in an extended vertical form in running water to remove residual solvent.   A minimum washing time of 17 hr for filaments spun from a solution in dimethyl acetamide and of 48 hr for filaments spun from a solution in 1-formylpiperidine or its mixtures was found to be necessary to reduce the residual solvent content below 3%. Figure 2 shows the rate of removal of the latter solvent by such washing, residual solvent being determined by thermogravimetric analysis.

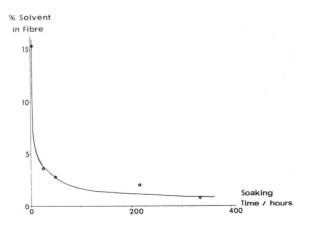

Figure 2.   Rate of removal of 1-formylpiperidine from
fibres by washing.

There is a serious problem of damage to the spinneret during handling
and particularly during cleaning.   If the injection tube is moved even
slightly off line it ceases to be concentric with the outer circumference of
the hole so concentric holes can no longer be obtained in the hollow filaments produced.   A solution to this problem has been proposed and illustrated
in a Monsanto patent.[12]   Adjustable screw probes are built in to the spinneret
with their points contacting the outside of the hollow tube, such that by
adjusting the probes the position of the tube can be altered to make it
central within the spinneret hole.

Gas permeability measurements

Fibres were tested in bundles of 10 fibres each of length 20 cm.   One
end of the bundle was sealed by epoxy resin.   The bundle was mounted in a
gas separation module and permeate flows from the open end were measured at
room temperature at a pressure differential of 100 psi from the outside of
the fibres to the gas exit from inside the fibres.

Separation factors (selectivities) were calculated from the ratio of
permeation rates of invididual gases, not from permeation data for mixtures.
The results given are for hydrogen against methane, but data for carbon
dioxide were also obtained and the general pattern of variation of permeation
rate and of selectivity against the other two gases was found to be similar.

Silicone coating

Coating was carried out using a solution of 2% Sylgard 184 (Dow Corning)
and 0.25% of the curing agent supplied in isopentane.   The fibre bundle,
mounted as for incorporation in a module, was evacuated from the inside while
immersed in the solution for 10 min., removed and dried while still evacuated
from within, then allowed to cure in air for at least a week before use.

Effect of production conditions on fibre properties

1.   Polymer extrusion rate

At constant water injection rate and wind-up speed, the effect of increasing polymer extrusion rate was, as expected, to increase the fibre
diameter (Figure 3) and wall thickness and hence to reduce the permeation
rate (Figure 4) and increase the selectivity (Figure 5).   At low extrusion
rates the fibre wall became so thin that it was easily deformed on contact
with the rollers in the coagulation bath so that a collapsed or tape-like
fibre was formed.   At very low extrusion rates the pressure of internal
coagulant burst the fibre wall and spinning became impossible.

There was a slight increase in void volume fraction with increase in polymer extrusion rate, which is attributable to a higher ratio of internal void structure to relatively dense surface structure in the thicker walls.

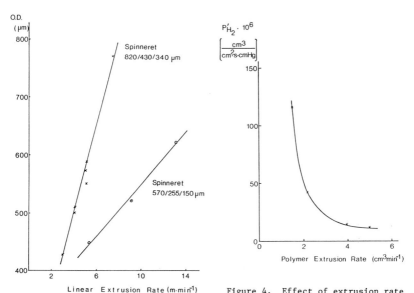

Figure 3. Effect of extrusion rate on fibre diameter (O.D.)

Figure 4. Effect of extrusion rate on permeation rate (P').

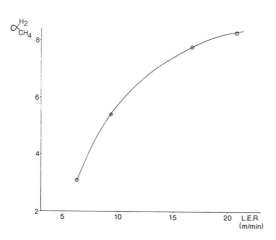

Figure 5. Effect of extrusion rate on selectivity ($\propto$).

2. Water injection rate

    At low water injection rates, slot-shaped holes were produced.   At
constant polymer extrusion rate and wind-up speed, increasing water injection
rate gave progressively rounder holes, with an increase in hole diameter and
fibre diameter and a reduction in wall thickness.   The ratio of fibre bore
area to total fibre cross-sectional area rose typically from 0.25 to 0.5
with a 5-fold increase in injection rate.   These ratios are higher than the
majority of those described in Monsanto patents.

Higher water injection
rate caused an increase in
the void volume fraction in
the fibre (Figure 6), presumably due to a higher proportion of macrovoids in the
thinner cells, and also led
to an increase in hydrogen
permeation rate (Figure 7)
and a fall in selectivity.

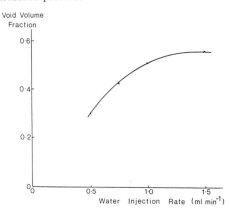

Figure 6.   Effect of water injection
rate on void volume fraction.

3. Air Gap

    Varying the air gap from 0.5 to
8.0 cm gave little or no change in
permeation rate or selectivity.

4. Coagulation bath temperature

    Varying the bath temperature
between 15 and 45°C gave no substantial change in permeation rate or
selectivity.

5. Polymer solution extrusion
    temperature

    Varying the extrusion temperature
between 20 and 45°C gave little or no
change in permeation rate or selectivity.

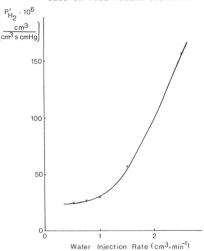

Figure 7.   Effect of water injection
rate on permeation rate (P').

6. Production rate

    When the wind-up speed is increased without increasing polymer extrusion
rate and water injection rate, the result is an increase in spin stretch
ratio and correspondingly lower internal and outer diameters with the expected
increase in permeability.   However, when the wind-up speed is increased with
corresponding increases in polymer extrusion rate and water injection rate,
the result is an increase in production rate of a fibre with virtually unchanged dimensions.   A range of production rates from 5 m min^{-1} to 30 m min^{-1}
gave an improvement in circularity as the rate rose but no systematic differences in permeation rate or selectivity.   Higher spinning speeds are
desirable for improved productivity but could not be attained on the equipment
in use.

7. Polymer concentration

    Increasing polymer concentration in the spinning solution from 32% to
38% decreased the void volume (Figure 8) and the proportion of macrovoids.
The hydrogen permeation rate was more than an order of magnitude higher at
32% than at 35% or 38% concentration;   selectivities did not differ appreciably from those expected for a gaseous diffusion process at 32% concentration
but were significantly higher at 35% and 38%.

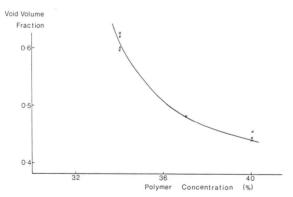

Figure 8.   Effect of polymer concentration on void volume
fraction.

8. Wall thickness

    As expected, permeation rates were lower for thicker walls (Figure 9).
The effect of wall thickness was greater than would be forecast if the major
resistance to permeation lay in invariant outer and inner skins.   Above a

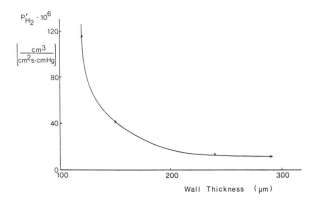

Figure 9.   Relation between fibre wall thickness and permeation
rate (P').

wall thickness of about 100 μm a further increase in wall thickness produced
a very marked increase in selectivity (Figure 10), but below this critical
value the selectivities were approximately those for separation by gaseous
diffusion.   These results can be explained if the skin porosities were higher
for the thinner walls or, less probably, if the resistance to permeation lies

mainly in the cell structure
rather than in the skin layers.

9.   Drawing

   Polysulphone hollow fibres
could be cold drawn to a draw
ratio of about 1.8.   The effect
of drawing at a draw ratio of
1.5 was to increase the tensile
strength by about 40% and the
Young's modulus by about 30% and
to reduce the elongation at break
from about 80% to about 25%.   The
gas permeation rate through the
fibre was greatly increased by
drawing, to a greater extent than
would be expected for the decrease
in wall thickness.   This result

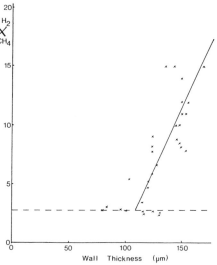

Figure 10.   Relation between fibre wall
thickness and selectivity (∝).

indicates that the increase in surface area resulting from drawing is accompanied by an increase in the frequency of channels penetrating the surface.

10. Solvent

Dimethyl acetamide tends to give fibres containing large microvoids with
their major axes perpendicular to and located just inside outer and inner
skin structures, with a relatively microporous structure in the centre of the
wall (Plate 2a).   Use of 1-formylpiperidine (FP) as a solvent gave fibres
that were relatively free from macrovoids (Plate 2b), and the further addition
of 5% or 10% of formamide (FA) to the 1-formylpiperidine reduced the macrovoid content still further (Plates 2c,d).   Macrovoid content with this
solvent system was low even when a solution strength as low as 30% was
extruded;   such a low polymer concentration in dimethyl acetamide would
accentuate macrovoid formation.   The FP/FA (90/10) solvent gave fibres
characterised by a smaller and more uniform pore size, thinner inner and
outer skins, and no macrovoids;   gas permeabilities in these fibres were very
high, and selectivities were low.

Collapse pressures

Collapse pressures were measured by increasing the pressure of nitrogen
on the outside of the
fibre and measuring
the resulting permeate
flow.   Figure 11 shows
typical collapse behav-
iour for fibres spun
from 35% solutions of
P3500 in dimethyl
acetamide.

Use of Victrex poly-
ether sulphone

Hollow porous fibres
were prepared using
Victrex 200P poly-

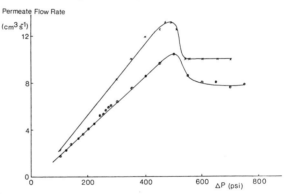

Figure 11.   Typical collapse pressure data for
fibres spun from 35% solutions in
dimethyl acetamide.

ether sulphone.   This polymer gave less viscous solutions than P3500, so that
spinning could readily be carried out near to room temperature (24.5°C) using
a more concentrated spinning dope containing 40% (w/w) of polymer in dimethyl
acetamide.   Data are given in Table 1.   These fibres had the lowest permeation rates and highest selectivities of all the uncoated fibres made in this
series.   However, the values obtained are consistent with extrapolation of
the data obtained for P3500 solutions of various concentrations to this higher
concentration and are not necessarily attributable to a significant difference
in chemical behaviour between the two types of polysulphone.

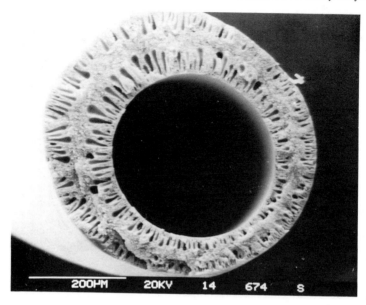

(a) dimethyl acetamide

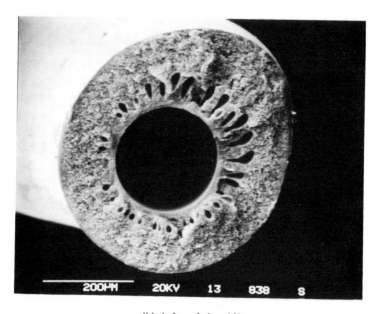

(b) 1-formylpiperidine

Plate 2. Fibres spun from various solvents

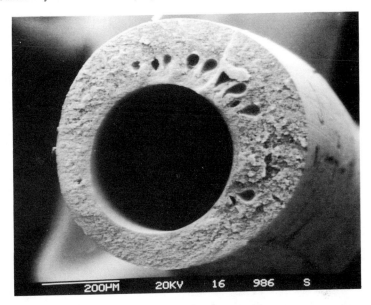

(c) 95/5 1-formylpiperidine/formamide

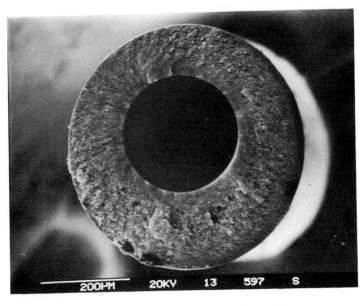

(d) 90/10 1-formylpiperidine/formamide

TABLE 1

Hollow porous fibres from Victrex 200P

Spinneret: 820/320/200   40% Victrex 200P in dimethyl acetamide
Polymer solution temp. 24.5°C   Coagulant temp. 21.5°C
Air gap: 7.5 cm $\underline{P}$ = peristaltic pump; $\underline{S}$ = syringe

Extrusion rate ml min^{-1}	Water injection rate ml min^{-1}	Wind-up speed m min^{-1}	Spin stretch factor	Fibre O.D. μm	Hydrogen permeation rate P'_{H_2} x 10^6	H$_2$ CH$_4$	CO$_2$ CH$_4$
2.2	1.5 $\underline{P}$	19.3	3.9	400	8.9	26	11
4.0	0.45 $\underline{P}$	14.3	1.6	460	3.8	28	12
9.1	11.0 $\underline{S}$	26.3	1.3	850	3.8	20	8.5

Effect of coating fibres

    Fibres coated with Sylgard 184 had lower gas permeation rates than un-
coated fibres, and in most cases also had higher selectivities.   Figure 12
summarises data for uncoated fibres and Figure 13 data for coated fibres.

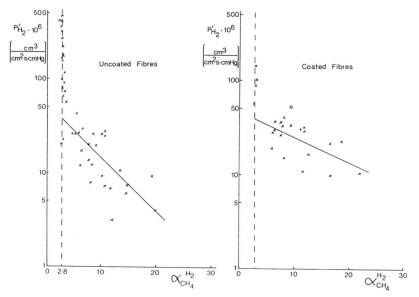

Figure 12. Relation between permeation   Figure 13. Relation between
            rate (P') and selectivity (α)   permeation rate (P') and
            for uncoated fibres.   selectivity (α) for coated
                                                         fibres.

It appears that for both sets of fibres the selectivity ratio for hydrogen against methane is approximately the theoretical value of 2.8, calculated for separation by gaseous diffusion alone, for permeation rates higher than about $40 \times 10^{-6} cm^3/cm^2$ s cm Hg, but as permeation rates decrease below this limit the selectivities progressively improve.   As shown in Figure 14, in this region selectivities are better at a given permeation rate for coated fibres than for uncoated fibres.

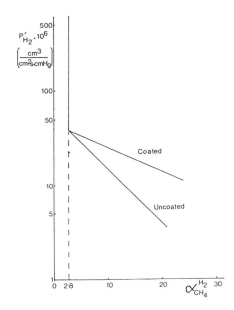

Figure 14.   Relations between permeation rates (P') and selectivities (α) for coated and uncoated fibres.

Conclusions

   1. For the range of uncoated and silicone-coated hollow porous poly-sulphone fibres described in this paper, selectivities for gas separations are higher the lower the gas permeation rates.
   2. Above a gas permeation rate of about 40×10^{-6} cm^3 cm^2 s^{-1} cm Hg^{-1} selectivities do not exceed the theoretical values for separation by a gaseous diffusion process.
   3. Below a gas permeation rate of about 40×10^{-6} cm^3 cm^{-2} s^{-1} cm Hg^{-1} selectivities are higher for silicone-coated than for uncoated fibres at the same gas permeation rate.
   4. For uncoated fibres made using dimethyl acetamide as solvent, the most important spinning parameter in terms of permeation rate and selectivity

is the polymer concentration in the spinning solution.   Higher polymer concentrations give lower permeation rates, higher selectivities, thicker inner
and outer skins, and less macrovoid formation.

5. Some other spinning parameters, such as lower polymer extrusion
rate, higher water injection rate and higher wind-up speed, also increase
permeation rates and reduce selectivities, partly at least through their
effect on wall thicknesses.

6. Fibres made using 1-formylpiperidine as solvent, especially in
association with minor amounts of formamide, have smaller and more uniform
pore size, thinner inner and outer skins, little or no macrovoid structure,
and higher gas permeation rates relative to those made using dimethyl
acetamide.

Acknowledgments

This work was supported by British Gas, London Research Station.   The
authors thank the staff of the Gas Properties and Measurement Group at L.R.S.
for their collaboration.

References

1. S.A. Stern, "The Separation of Gases by Selective Permeation", Chapter 8
    in P. Meares (ed.), Membrane Separation Processes, Elsevier, 1976,
    pp. 295-326.

2. J.M.S. Henis and M.K. Tripodi, Science, 1983, 220, 1 April, 11.

3. Monsanto Co. (J.M.S. Henis and M.K. Tripodi), U.S. Patent 4,230,463
    (pub. 28 Oct., 1980).

4. Daicel Ltd. (K. Arisaka, K. Watanabe and K. Sawazima), U.S. Patent
    4,127,625 (pub. 28 Nov., 1978).

5. Monsanto Co. (R.R. Ward, R.C. Cheng, J.C. Danos and J.A. Carden),
    U.S. Patent 4,214,020 (pub. 22 July, 1980).

6. Monsanto Co. (A.A. Brooks, J.M.S. Henis and M.K. Tripodi), U.S. Patent
    4,364,759 (pub. 21 Dec., 1982).

7. J.M.S. Henis and M.K. Tripodi, J. Membrane Sci., 1981, 8, 233.

8. H.K. Lonsdale, J. Membrane Sci., 1982, 10, 81.

9. I. Cabasso, E. Klein and J.K. Smith, J. Appl. Polymer Sci., 1976, 20,
    2377.

10. S. Yamamoto, H. Ujigawa, S. Ohnishi, O. Hirasa, M. Kato and M. Kusumoto,
    Bull. Res. Inst. Polym. Text., 1984, 12, No. 144, 101.

11. Monsanto Co. (A.A. Brooks, J.M.S. Henis, J.E. Kurz, M.C. Readling and
    M.K. Tripodi), U.K. Patent Appl. GB 2,047,162A (pub. 26 Nov., 1980).

12. Monsanto Co. (F.A. Goffe), U.S. Patent 4,493,629 (pub. 15 Jan., 1985).

Hollow Fiber Gas Separation Membranes: Structure and Properties

By Z. Borneman, J. A. van t'Hof, C. A. Smolders*, and H. M. van Veen

DEPARTMENT OF CHEMICAL TECHNOLOGY, TWENTE UNIVERSITY, PO BOX 217, 7500 AE ENSCHEDE, THE NETHERLANDS

Summary

The structure and properties of asymmetric polyethersulfon (PES) hollow fiber membranes have been investigated in relation to the spinning conditions.

The hollow fibers were prepared by a dry-wet spinning process from a solution of PES, a polyalcohol and NMP. The most important parameters during the spinning process are the composition and temperature of the spinning dope and the coagulation bath, the dimensions of the spinneret used, the length of the airgap (distance between spinneret and external coagulation bath) and the spinning rate.

The fibers have been tested on flux and selectivity for a carbon dioxide/methane-mixture on a laboratory test-facility. Characterization of the structure was done by a scanning electron microscope.

It was found that the membrane selectivity increased upon decreasing the temperature of the external bath. Below 20 °C the permeability reached a constant value, while the selectivity still increased.

A higher flux value with hardly any change in selectivity was obtained by adding a polyalcohol to the internal bath. It was also observed that the number and size of macrovoids at the bore side of the fiberwall reduced. At high polyalcohol concentration (more than 25%) small voids appeared at the shellside of the fiber.

Another important variable was the temperature of the internal (bore side) nonsolvent. By raising this temperature fibers were obtained with a better permeability, while the influence on the selectivity remained small.

Silicone-coated fibers showed an increase in both selectivity and flux when polyalcohol was added to the internal bath during spinning or when the temperature of this bath was raised. This is remarkable because in most cases an improvement in selectivity is accompanied by a loss in permeability.

Introduction

During the last 10 years a lot of research has been done in the field of
gas separation with membranes [1]. The fact that membrane processes are
simple in operation (no moving parts, no regeneration steps), do not need
excessive amounts of energy (compared to cryogenic processes), and use little space, make them very interesting for commercial use.

A limiting factor of gas separation membranes is the insufficient selectivity in combination with a high flux. From literature it is known that some
polymers have very good selectivities but extremely low flux values ('bar-
rier-polymers' like polyvinylalcohol and polyacrylonitril). On the other
hand there are also polymers with very high fluxes and low selectivities
(elastomeric polymers like silicone rubber) [2].

Polyethersulfone (PES) is an example of a polymer with a rather good selectivity (for a carbon dioxide/methane-mixture a value of 40 has been reported [3]) at moderate permeabilities (for CO_2 about 3 Barrer). In order to
use this material commercially in gas separation, the membranes should be
very thin but dense (the flux through a membrane is inversely proportional
to the membrane thickness) and the modules should have a large ratio of
membrane area to overall volume. This can be realised by using hollow fiber
membranes (high area to volume ratio) with an asymmetric structure (a thin
selective layer, supported by a porous substructure).

Asymmetric hollow fibers can be spun by a dry-wet spinning process [4].
This spinning process is shown schematically in figure 1. A homogeneous
polymer solution is extruded through a tube-in-orifice spinneret (orifice
dimensions smaller than 1 mm). After a short airgap the fiber is immersed
into and transported through a nonsolvent bath and collected. Parameters of
interest during this spinning process are: the composition and temperature
of the polymer solution, of the bore liquid (internal bath) and of the external bath, the length of the airgap and the spinning rate.

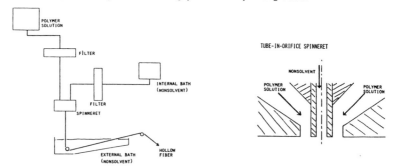

Figure 1. Schematic representation of the dry-wet spinning process.

In our laboratory it was found that membranes with a fine porous substructure (without many 'macrovoids') could be made from a 30% solution of polyethersulfone (PES) in NMP to which some polyalcohol (5-15%) was added. The
effect of the polyalcohol on the wall structure is shown in figure 2. The
selectivity of such membranes was very low: about 1-3. To find out whether
the intrinsic separation properties of the polymer could be achieved, the
fibers were coated with a thin layer of silicone rubber to block (small)
pores present on the fiber surface. This coating appeared to be very effective and selectivities of 25 to 30 were easily obtained.

 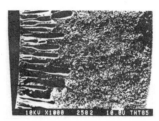

Figure 2: Electron micrographs of PES fibers with 0% (A) and 5%(B) polyalcohol in the spinning solution.

This paper describes the results of a systematic investigation on the
influence of the following parameters on the structure and properties of
fibers spun out of a solution of 30% PES, 5% of a polyalcohol and 65% NMP at
30 °C:

1) the composition and temperature of the internal coagulation bath;
2) the temperature of the exteral bath.

The investigations were done by measuring the permeability and selectivity
of the fibers for a mixture of carbon dioxide and methane and by examining
the membrane structure with a scanning electron microscope.

Theory

The separation of gases through dense polymeric membranes is based on a
difference in solubility and diffusivity. The gasflux Q [volume/time]
through such membranes is given by the following equation:

$$Q = P.A.\Delta p/L \qquad (1)$$

where P is the permeability of a gas through the polymer, A the membrane area, Δp the pressure difference over the membrane and L the membrane thickness.

In order to have good separation characteristics, the membrane must have a high flux as well as a high selectivity for one of the components. The high selectivity is mainly determined by the difference in permeabilities P_i for the components i, while the flux is also determined by the membrane geometry and the way of operation, as can be seen from formula (1).

To achieve a flux as high as possible for a given polymer, the membrane area A should be large and the membrane thickness L small. This can be realised by using asymmetric hollow fibers. This type of membrane geometry is very suitable for processes with gases, where contamination is just a minor problem.

Immersion precipitation is a widely used technique to obtain asymmetric membranes: here, a polymer solution (polymer and solvent) is cast on a support as a thin film and immersed in a nonsolvent bath. The polymer precipitates because of an exchange of solvent and nonsolvent.

In figure 3 a schematic representation is given of a ternary phase diagram of polymer, solvent and nonsolvent In this diagram three regions can be distinguished: (1) a one-phase region, where the three components are completely miscible with each other; (2) a two-phase region, where two coexisting liquid phases (a polymer-rich phase and a polymer-poor phase) are formed and (3) a region in which the formation of aggregates and small crystalline entities takes place, accompanied by gelation of the solution.

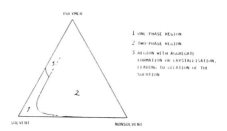

Figure 3:   Ternary phase diagram of the system polymer-solvent-nonsolvent.

Formation of the top layer of an asymmetric membrane occurs by gelation, and the structure (density) of this layer is mainly determined by the polymer concentration at the interface with the coagulation bath. The porous sublayer is obtained by liquid-liquid demixing through nucleation and growth of the dilute phase. The growth of the droplets is limited by gelation of the surrounding polymer-rich phase. Partial coalescence of the droplets gives a structure of connected pores. Here the interaction between solvent and nonsolvent is one of the main factors determining the structure.

The formation of hollow fiber membranes is more complex, because here the polymer solution is not cast as a film but extruded through a narrow orifice, where precipitation of the polymer starts from two sides: from the inside (the bore) of the fiber and from the outside (the external bath). Both 'coagulaton fronts' influence each other and can therefore have a large impact on the membrane structure.

Furthermore, adding another component (e.g. a poly-alcohol) to the polymer solution and/or the coagulation bath, the theoretical description of membrane formation becomes more complex. In a forthcoming paper the mechanism of membrane formation will be described in more detail [5].

Experimental

The polymer used was polyethersulfon VICTREX[R] from ICI and the solvent N-methyl-2-pyrrolodine (NMP) was of analytical grade. The polymer was dried at elevated temperatures (3-4 hours at 80 °C) and the solvent and additive were used without further pre-treatment. De-ionised and hyperfiltrated water was used as a nonsolvent.

The spinning solution had a temperature of 30 °C and the fibers were spun on a laboratory-scale spinning apparatus with a speed of about 10 m/min. afterwards the fibers were rinsed with tapwater for 4-6 hours and dried at room temperature for 1-2 days at a relative humidity of 50%.

For testing purposes modules of 10-15 fibers were made and tested at room-temperature with a feed of 15-50 vol-% CO_2 in methane at a pressure of 3-5 atm. The permeate side of the fibers was evacuated and by measuring the pressure increase on the permeate side during a certain time in a well-known volume, the permeability per unit membrane thickness (P/L) could be calculated. A schematical drawing of the experimental set-up is given in figure 4.

Afterwards the fibers were coated with a layer of silicone rubber (Sylgard 184, Dow Corning) and the permeation through these fibers was also measured.

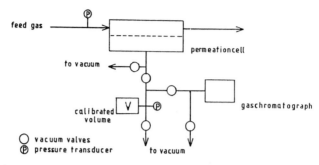

Figure 4:   Equipment for gaspermeabilty measurements.

To investigate the membrane structure, the fibers were prepared by cryogenic breaking and sputtered with a small layer of gold. The cross-sections
of the fibers were examined with a Jeol 35 CF scanning electron microscope.

Results and discussion

In this section the results of the gas separation measurements are given,
together with some of the electron micrographs.

Polyalcohol in the internal bath. From literature it is known that adding a
polyalcohol to the coagulation bath gives the membranes a more porous structure, with a reduction of the number of macrovoids [6]. Such a structure is
very important when the membranes have to be used at higher pressures.

To see whether such a porous structure would also result in a higher permeability, different amounts of the polyalcohol were added to the internal
coagulation bath. The results of the permeability measurements are given in
table 1.

The results in table 1 show that adding polyalcohol to the internal bath
gives an increase in permeability for the uncoated fibers and only a slight
change in selectivity.

For the coated fibers in all cases a much higher selectivity is obtained
together with a decrease in permeability. When a polyalcohol is added to the
internal bath both flux and selectivity become better upon coating.

Especially this latter effect is striking. The scanning electron
micrographs (figure 4) show that the macrovoids on the inside are getting
somewhat smaller when a polyalcohol is added.

Table 1:   Influence of polyalcohol in the internal coagulation bath

internal bath (w-% polyalcohol)	gasflux $(P/L)_{CO_2}$ $(cm^3/cm^2.s.cmHg)10^6$	selectivity CO_2/CH_4	
0	14.7 (8.0)	1.0 (16.2)	I
10	21.6 (9.4)	0.9 (20.0)	
0	17.5 (9.7)	1.3 (20.4)	II
25	22.0 (11.8)	1.3 (29.6)	
0	7.4 (5.3)	3.3 (19.0)	III
25	26.7 (5.5)	1.0 (25.3)	

(..): values after coating with silicone rubber
external bath: 100% water, T = 20 °C
internal bath: T = 20 °C
spinning rates are different for the 3 types of fibers I to III.

Figure 5:   Electron micrographs of PES fibers, spun with 0% (A) and 25% (B) polyalcohol in the internal bath.

The fine-porous structure of the fiber wall seems to have a better gas permeability than the structure with larger macrovoids. The influence of a toplayer on the inside of the membranes might affect the membrane performance but this cannot be deduced from the electron micrographs (figure 6).

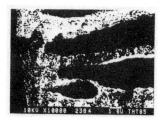 

Figure 6: Electron micrographs of PES fibers, spun with 0% (A) and 25% (B)
polyalcohol in the internal bath (detail inner skin).

An explanation for the fact that the selectivity of coated membranes increases with more polyalcohol in the bore fluid during spinning can be given
when the 'resistance model' of Henis et al. [7,8] is used. In this model it
is stated that when the sublayer of an asymmetric membrane contributes to
the resistance for the gas which passes through the membrane, this effect is
bigger for the selectively permeating gases (like CO_2) than for the 'slow'
permeating gases (e.g. CH_4), which results in a rather low selectivity after
coating. With the polyalcohol in the internal bath it is assumed that the
resistance of the sublayer is lowered which gives a better selectivity of
the coated fibers.

Temperature of the internal bath. In the case of flat membranes it is known
that a higher temperature of the coagulation bath results in a more open
structure [9]. The reason is that with increasing temperature the diffusion
of nonsolvent from the coagulation bath to the polymer film is much faster,
which means that demixing takes place at a lower polymer concentration giving a more open structure. Another effect is that the toplayer becomes much
thinner.

The results of this temperature effect are given in table 2 and illustrated in figure 7.

As can be seen from table 2 and figure 7, for the uncoated fibers the
permeability is getting better when the temperature of the bath is raised,
while the selectivity remains very low. Bringing a layer of silicone rubber
onto the fibers a better selectivity is obtained, but the values are not as
high as in table 1.

Table 2:   Influence of the temperature of the internal coagulation bath

temp. internal bath (°C)	gasflux $(P/L)_{CO_2}$ $(cm^3/cm^2.s.cmHg)10^6$		selectivity (CO_2/CH_4)	
20.5	23.2	(11.9)	1.1	(7.9)
28	32.4	(14.8)	1.0	(16.4)
38	62.1	(19.5)	1.0	(18.7)
47	69.2	(18.5)	1.0	(19.9)

(..): values after coating with silicone rubber
external bath: 100% water, T = 20 °C
internal bath: 100% water

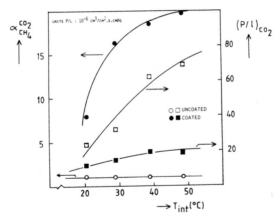

Figure 7:   Selectivity and flux as a function of the internal bath temperature.

For the coated fibers both flux and selectivity increase with increasing
temperature of the internal bath which is probably due to a decreasing resistance of the sublayer.

Figure 8 shows the structures of membranes spun at 20 °C (A) and 47 °C

(B). At the higher temperature small macrovoids are appearing at the shell-side of the fibers.

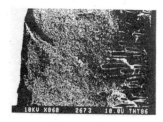 

Figure 8: Electron micrographs of PES fibers, spun at various temperatures of the internal bath: 20 °C (A) and 47 °C (B).

Temperature of the external bath. In order to get a more dense toplayer on the shellside of the membrane, the polymer concentration at the moment of gelation should be as high as possible [9]. This can be done by removing solvent at the interface with the external bath, without a large instream of nonsolvent. A way to achieve this is cooling the external bath, so that the diffusivity is much smaller. To see the effect of the temperature the following experiments have been performed (table 3 and figure 9).

Table 3:   Influence of the temperature of the external coagulation bath

temp. external bath (°C)	gasflux $(P/L)_{CO_2}$ $(cm^3/cm^2.s.cmHg)10^6$		selectivity CO_2/CH_4	
41	37.2	(18.5)	1.1	(2.8)
30	27.0	(12.5)	1.3	(7.0)
21	22.3	(10.1)	1.3	(17.1)
14	17.2	(10.9)	1.6	(23.3)

(..): values after coating with silicone rubber
internal bath: 100% water, T = 20 °C
external bath: 100% water

From table 3 it is clear that the temperature of the external bath has a big influence on the membrane properties. The selectivity of the coated membranes raises to a value of more than 20 when the external bath is cooled

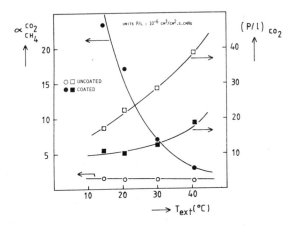

Figure 9: Selectivity and flux as a function of the external bath temperature.

and it seems that a still further lowered temperature could give an even better selectivity. The change in permeability is not so drastic and seems to reach a constant value of about $10-11.10^{-6} cm^3/cm^2.s.cmHg$.

Figure 10: Electron micrographs of PES fibers, spun at various external bath temperatures: 40 °C (A) and 14 °C (B).

The structures of fibers spun with a high temperature and a low temperature of the external bath are given in figure 10. Here again the reduction
of the length of the macrovoids can be seen with decreasing temperature.

Conclusions

From the experiments shown in the previous section it is clear that the
investigated variables have an important influence on the membrane properties.

With a solution of 30% polymer and 5% polyalcohol in NMP fibers can be
spun with a high flux, but with a low selectivity. Taking the right values
of the spinning parameters and coating the fibers, good selectivities can be
obtained for CO_2/CH_4 (about 30) with still an acceptable flux (P/L value)
for carbon dioxide (P/L = 12.10^{-6} $cm^3/cm^2.s.cmHg$).

By adding polyalcohol to the internal coagulation bath or by raising the
temperature of this bath, fibers are obtained with both a higher selectivity
and a higher flux after coating.

Of course there are more parameters which determine the structure and properties of the membranes (e.g. the dope composition and the spinning rate)
and we hope to publish on that matter in the near future.

To make selective membranes without a coating, other spinning conditions
have to be used. The results of the experiments described in this paper can
be used to optimize the selectivity and permeability of such a fiber with
good intrinsic properties.

Acknowledgement

This work is sponsored by the Ministry of Economical Affairs in the
Netherlands as an IOP-project on membranes.

References

[1] a. V.T. Stannet et al., Adv. Pol. Sci., 1979, 32, 69.
    b. S.L. Matson et al., Chem. Eng. Sci., 1983, 38, 503.
    c. H. Finken and Th. Krätzig, Chemie-Technik, 1984, 13(8), 75.

[2] S.S. Kulkarni et al., AIChE Symp. Ser., 1984, 79(229), 172.

[3] D.E. Ellig et al., J. Membr. Sci., 1980, 6, 259.

[4] a. I Cabasso et al., in: "Kirk-Othmers Encyclopedia of Chemical Technology", 1980, 12, 492.

b. P. Aptel et al., J. Membr. Sci., 1985, 22, 199.

[5] J.A. van 't Hof et al., to be published.

[6] H. Strathmann, "Trennung von molekularen Mischungen mit Hilfe synthetischer Membranen", Darmstadt, 1979.

[7] J.M.S. Henis and M.K. Tripodi, J. Membr. Sci., 1981, 8, 233.

[8] F.W. Altena, private communication.

[9] L. Broens, F.W.Altena, C.A. Smolders and D. Koenhen, Desalination, 1980, 32, 33.

Structure-Permeability Relationships in Silicone Polymer Membranes*

By S. A. Stern**, V. M. Shah, and B. J. Hardy

DEPARTMENT OF CHEMICAL ENGINEERING AND MATERIALS SCIENCE, SYRACUSE
UNIVERSITY, SYRACUSE NY 13244, USA

Remarkable progress has been made during the last two decades in the
technology of gas separation by selective permeation through nonporous polymer
membranes. Thus, up to the middle 1960's this technology was being studied
only in the laboratory or on pilot-plant scale. At present, by contrast, scores
of permeation plants are separating various gas mixtures on an industrial scale
in the U.S. and in several other countries. Some of these plants process gas
streams as large as 1×10^6 to 5×10^7 cu. ft./day.

In order to sustain the progress in membrane separation technology in
future years, it will be necessary to develop an ability to "tailor-make" polymer membranes for desired gas separation processes. In other words, it will be
necessary to synthesize polymers that exhibit both a high <u>intrinsic</u> permeability
and a high selectivity toward specific components of gas mixtures of interest.
Progress in this area is greatly impeded at this time by a lack of sufficient
understanding of the relationships between the chemical structure of polymers
and their gas permeability. The present study was undertaken in order to gain
an insight in these relationships. Silicone polymer membranes were selected
for study because of their relatively high gas permeability, and also because
the versatility of silicone chemistry is such that a large variety of different
functional groups can be substituted in the polymer backbones or side chains.

*To be published in J. Polym. Sci., Polym. Phys. Ed.
**To whom correspondence should be addressed.

158

Accordingly, permeability coefficients, $\bar{P}$, for He, O_2, N_2, CO_2, CH_4, C_2H_4, C_2H_6, and C_3H_8 in 12 different silicone polymer membranes were studied at 35.0°C and pressures up to 9 atm. The polymers were synthesized by Dr. C.-L. Lee and his coworkers at Dow Corning Corp. of Midland, Michigan. Values of $\bar{P}$ for CO_2, CH_4, and C_3H_8 were also determined at 10.0 and 55.0°C. In addition, mean diffusion coefficients, $\bar{D}$, and solubility coefficients, S, were obtained for CO_2, CH_4, and C_3H_8 in 6 silicone polymers at 10.0, 35.0, and 55.0°C. Solubility and diffusion coefficients were required for the interpretation of the experimental data because gas permeation through nonporous polymer membranes is known to occur by a "solution-diffusion" mechanism. It can be shown that $\bar{P} = \bar{D}.S$.

Substitution of increasingly bulkier functional groups in the side- and backbone-chains of silicone polymers results in a significant decrease in $\bar{P}$ for a given penetrant gas. This is due mainly to a decrease in $\bar{D}$, while S decreases to a much lesser extent. Backbone substitutions appear to have a somewhat lesser effect in depressing $\bar{P}$ than equivalent side-chain substitutions.

The selectivity of a silicone membrane for a gas A relative to a gas B, i.e. the permeability ratio or "ideal" separation factor $\alpha^* \equiv \bar{P}(A)/\bar{P}(B)$, may increase or decrease as a result of such substitutions, but only if the substituted groups are sufficiently bulky. The selectivity of the more highly permeable silicone membranes is controlled by the solubility ratio S(A)/S(B), while the selectivity of the less permeable membranes depends on both the diffusivity ratio $\bar{D}(A)/\bar{D}(B)$ and the solubility ratio S(A)/S(B). These ratios are known as the "mobility selectivity" and the "solubility selectivity", respectively.

The selectivity of one silicone polymer , $[(CF_3CH_2CH_2)CH_3SiO]_x$, toward CO_2, relative to other gases, was found to be significantly higher than that of the analog polymer which did not contain fluorine, namely $(C_3H_7CH_3SiO)_x$. The enhanced CO_2 selectivity of the former polymer probably is due to specific interactions between its polar, fluorine-containing side groups and CO_2. This suggests that the selectivity of other polymers toward CO_2, and possibly toward O_2, could also be enhanced by the substitution of functional groups which would induce specific interactions with these gases.

Thomas Graham and Gaseous Diffusion

By M. Stanley

MANOR HIGH SCHOOL, WEDNESBURY, WEST MIDLANDS, UK

In order to appreciate Graham's work on gaseous diffusion I propose to begin with a review of earlier studies. Graham wrote of diffusion that: "the earliest observations we possess on this subject are those of Dr. Priestley."[1] Beginning in 1777, Joseph Priestley mixed gases such as carbon dioxide and common air and discovered that they did not appear to separate out into distinct strata on standing, rather they remained uniformly diffused. However, he added a caveat that if two gases were placed carefully in a vessel without any mixing, then "they might separate, as with the same care water and wine may be made to do; but that when once they have been mixed, they will continue to be so, like wine and water after having been shaken together."[2]

In 1783, Priestley encountered diffusion again in another context. James Watt had suggested to him that it might be possible to make 'permanent' steam by converting all the latent heat of steam into sensible heat. In an attempt to achieve this conversion Priestley heated wet lime in a clay retort and collected the gaseous product over mercury. To his surprise, the product was not steam; it was common air.[3] The steam had been exchanged for air through the pores of the clay retort. Eventually, in 1786, he explained that the air had entered the clay pores by capillary attraction, but then added "in what manner it acts in this case, I am far from being able to explain. Much less can I imagine how air should pass one way, and vapour the other, in the same pores, and how the transmission of the one should be necessary to the transmission of the other."[4]

161

Priestley concluded his diffusion experiments in 1800 by showing that gases exchanged places with one another regardless of affinity. He enclosed different gases in earthenware tubes, which were closed at one end and inverted over water. By heating the top end of the tube he found that gases inside the tube such as hydrogen, oxygen, or nitrous air were exchanged for the common air outside. Significantly, he also observed that there was always a change in volume during this process but added, once again, "on what principle this change was made, I could not satisfy myself."[5]

Later, John Dalton attempted to explain diffusion by his first theory of mixed gases. According to this theory, if two gases A and B were mixed together then the particles of gas A would only repel those of their own kind. They would therefore not repel particles of gas B. In other words, particles of gas A were inelastic to those of gas B.[6] It was Newton who, in 1687, had originally suggested that gases might contain particles which repelled one another with a force inversely proportional to their distance apart.[7] However, he did not postulate as Dalton did the existence of selective forces of repulsion. Dalton believed that his own experiments of 1803 lent support to his first theory.[8] He experimented by connecting two glass flasks together using corks and a glass tube, having first put hydrogen into the upper flask and carbon dioxide into the lower one. After 30 minutes he found that the gases had spontaneously diffused into one another. Other gases were shown to interdiffuse, so Priestley's caveat was disproved, and Dalton was able to explain Priestley's exchange of gases through clay pores as being a diffusion process analogous to his two-flask experiment.

In France, Berthollet rejected Dalton's mechanical theory, that is the first theory of mixed gases, and supported instead, the alternative chemical solution theory, whereby gases would dissolve in one another by the action of a weak force of chemical attraction operating between different gases.[9] Likewise, the Scottish lecturers on Chemistry, Thomas Thomson and Thomas Charles Hope, were also opposed to Dalton's mechanical theory.[10] Hope complained that if Dalton was correct, diffusion should take only a few

seconds but it invariably took much longer. To overcome this difficulty

Dalton suggested that heat or caloric atmospheres, surrounding gas particles,

were the cause of the slow intermixture.[11] Hope responded by saying that

this was the strongest argument against Dalton's view. He wrote to Dalton

"this heat atmosphere I deem the essential cause of the elasticity and

repulsion among the particles of the gas, and I cannot conceive that this

atmosphere as it surrounds a particle of oxygen should repel the atmosphere

that surrounds another particle of oxygen and should not repel the atmosphere

that envelopes a particle of azote, hydrogen, or any other gas."[12]

   This comment went to the heart of the matter, as Dalton later conceded,

and in 1804 he responded to his critics by proposing an experimental test:-

it was to measure the speeds at which gases diffused into one another.[13] If

the speeds of diffusion differed,then he argued that affinity must be involved

and the chemical theory of diffusion would be favoured, but, if gases diffused

into one another at nearly the same speed then the mechanical theory would be

favoured. Surprisingly, Dalton did not observe any significant differences in

the values of the speed of diffusion of the same gas into a number of other

gases. In another experiment, designed to refute the chemical attraction

theory, he took a narrow tube, (12" x 0.25"), containing hydrogen, and found

that the loss of gas, in a given time, was unaffected by position, regardless

of whether the tube was pointing upwards, downwards or horizontally. These

experiments in support of the mechanical theory did not however convince his

critics.[14]

   Dalton therefore proposed a second theory of mixed gases which was

published in his New System of Chemical Philosophy of 1808. He now accepted

that all gas particles repelled one another. This was because each gas

particle consisted of an impenetrable nucleus, surrounded by a repulsive

atmosphere of heat.[15] There was now one force of repulsion - caloric. He

assumed further that all the particles inside a particular gas were identical

in size but differed in size from those of any other gas. Thus, he
envisaged a gas as an ordered or statical arrangement of particles held apart
by calorific repulsion.

Dalton then explained diffusion of the second theory by stating that when
two different gas surfaces met, there would be an unstable equilibrium owing to
the unequal sizes of the different gas particles. This unstable equilibrium
would produce an "intestine motion"[16] and so the particles of one gas would be
propelled amongst the others until a new equilibrium was achieved when each gas
was uniformly diffused through the other.

This theory did not entirely suit; for, in 1810, Dalton calculated that
five gases had particles possessing identical diameters.[17] Thomas Thomson
also expressed some doubts about the second theory and so he asked Dalton to
produce a mathematical demonstration of the motions produced when different-
sized gas particles were brought into contact.[18]

Further theoretical problems were raised by Berthollet's carefully-
performed two-flask diffusion experiments of 1809. Thomson wrote to inform
Dalton, that Berthollet had shown that "gases were always uniformly mixed in
24 hours if one of them was hydrogen.......; but other gases did not mix
uniformly in that time...... Air and carbonic acid $[CO_2]$ did not mix
uniformly in 17 days. In the highest globe there were 42 of carbonic acid,
in the lowest 50." [Thomson concluded characteristically] "These experiments
are rather adverse to your peculiar opinion respecting the gases."[19]

By February 1826, Dalton had reverted to his first theory of mixed gases
in a paper read to the Royal Society of London. In this paper on the
constitution of the atmosphere he argued from the first theory that there
should be relatively less oxygen at higher altitudes because oxygen possessed
a greater specific gravity than nitrogen.[20]

Thomas Graham, in his laboratory notebook of 1827-28, recorded that
"Dalton's theory of the tendency of gases to mix with each other, that they
are vacua to each other, is faulty, as it would occasion an instantaneous and

complete mixture, which certainly did not take place, and I have shown

previously that cold should be produced on mixture as a necessary effect of

Dalton's supposition, which I have found on trial not to be the case. The

gases must, therefore, actually press against each other."[21]

In his first paper on gaseous diffusion[22] Graham acknowledged that Dalton

and Berthollet had shown that two gases in contact diffuse spontaneously to

give a uniform mixture. He then described his own experiments to find the

speeds of diffusion of different gases into air. He used simple apparatus:-

a glass tube, 9" by 0.9", graduated into 150 divisions, and fitted with a

ground glass stopper containing a short, right-angled exit tube. From this

tube he allowed gases to diffuse into the air either upwards if they were

heavier than air or downwards if they were lighter than air. After ten hours

he analysed the residual gases in the tube and discovered that as much

hydrogen escaped in <u>two</u> hours as carbon dioxide did in <u>ten</u> hours. From this

Graham concluded "it is evident that the diffusiveness of the gases is

inversely as some function of their density - apparently the square root of

their density."[23] This statement of September 1829 was Graham's first

tentative announcement of his diffusion law.

Other important aspects of diffusion were revealed in this paper. Graham

had compared the speed of diffusion of a gas on its own, with its speed of

diffusion from a mixture of gases. He found that the more diffusive gas,

hydrogen, escaped in a greater proportion from a mixture of hydrogen and

ethylene than it did on its own. Thus began Graham's work on the diffusive

separation of gaseous mixtures which he later called 'atmolysis' - one of his

less successful neologisms. These experiments convinced him that diffusion

must involve the ultimate particles of gases and not sensible masses.

Dalton's friend, William Henry, in 1829, included copious extracts from

Graham's paper in the 11th edition of his textbook. He wrote that "Graham

has judiciously refrained, in the present state of the inquiry, from drawing

conclusions in favour either of the mechanical or chemical theory of the

constitution of mixed gases; but every attentive reader will, I think, perceive
that the tendency of the facts, so far as yet appears, is much in favour of
the mechanical explanation."[24]

In October 1830 Graham submitted an 18-page probationary essay to the
Faculty of Physicians and Surgeons of Glasgow so that he could become a
member of the Faculty in order to be recognised as a teacher of Chemistry in a
Scottish University. His essay was entitled:- 'On the tendency of air and
the different gases to mutual penetration'[25] and it reveals how Graham
gradually came to understand the nature of gaseous diffusion. He began by
describing how he had kept a vertical glass tube containing air for some
months and then analysed it. He found that the composition of the air was
identical at the top and bottom of the tube. The composition did not vary with
height as Dalton had suggested.

He then examined the current theories of diffusion. Firstly, he considered
Berthollet's chemical theory, that a weak force of chemical attraction caused
diffusion. He rejected this theory by arguing that"it is extremely unlikely
that the intensity of that attraction, and the consequent rapidity of the
diffusion, should depend entirely on the density of the gas, as we have found
to be the case."[25]

Secondly, Graham turned to Dalton's theory that gases spread or expanded
into one another as they would rush into a vacuum or at least, as he put it,
"that the resistance which the particles of one gas offer to those of another
is of a very imperfect kind, to be compared to the resistance which stones in
the channel of a stream oppose to the flow of running water."[25] After giving
details of his 1829-diffusion experiments Graham wrote that his own experiments
appeared to be compatible with the theory of Dalton.[25] However, further
experiments performed in the following year led him to change his mind.

A third alternative theory was that the diffusion of gases was sufficiently
accounted for by the elasticity and extreme mobility of gases. This was
essentially the position of Dalton's second theory and also latterly, that

of Graham's teacher, Thomas Thomson, but not that of Graham himself, who

wrote that:- "upon entering this enquiry, I very soon found that this last

supposition is altogether inadmissible, and that the diffusion of gases is

not of an accidental nature, but subject to fixed laws."[25]

Graham always claimed that the experiments of Johann Döbereiner were the

starting point for his own investigations into diffusion.[26] Interestingly,

Döbereiner had himself been anticipated by Priestley, who, in 1800, had noted

that broken glass vessels repaired with paint or cement allowed an exchange of

gases with the external air and that cracked vessels themselves were not

entirely impervious.[27] In 1823, Döbereiner had noticed that the water level

in a jar of hydrogen rose by $1\frac{1}{2}$" in 12 hours and by $2\frac{2}{3}$" in 24 hours.[28] The

apparent cause of this phenomenon was the escape of hydrogen through an

extremely minute fissure in the glass, and he showed that it could be prevented

if either oxygen, nitrogen,or air was used in place of hydrogen, or if the

cracked jar containing hydrogen was covered with a receiver. He explained that

the loss of hydrogen through the fissure occurred by capillary attraction,

adding that, only hydrogen atoms were small enough to penetrate the fissure

even although they were surrounded by the largest atmospheres of heat.

Graham regarded Döbereiner's explanation as improbable. He suggested

instead, that the rise in the water level was caused by an unequal diffusion

of hydrogen and air through the fissure - hydrogen being 3 or 4 times more

diffusive than air. Percipiently, he recognised that the process was

complicated, by differences between the inner and outer water levels, which

caused a pressure flow in addition to diffusion. By weighting the jar Graham

overcame this inequality and showed that hydrogen was about 4 times as

diffusive as air. Here, Graham established the important experimental

condition that gaseous diffusion must be studied at constant pressure. He

then showed that gases lighter than air, such as hydrogen or methane, made the

water level rise in a cracked jar, whereas ethylene, which had the same density

as air, gave no change in the water level, and heavier gases such as carbon

dioxide produced a slight fall. Graham concluded his Faculty paper by
pointing out that slightly damp Wedgwood unglazed, stoneware tubes were
better for demonstrating diffusion than cracked jars.

There was no mention of the use of stucco-in-glass tubes for diffusion
experiments or of the diffusion of gases into a vacuum, in the Faculty paper.
These experiments were therefore probably carried out after October 1830
because they were described in Graham's next paper on the law of the diffusion
of gases read to the Royal Society of Edinburgh on the 19th of December 1831.

In this paper Graham aimed to establish the numerical exactness of the
diffusion law which he stated at the outset as follows:- "the diffusion or
spontaneous intermixture of two gases in contact is effected by an interchange
in position of indefinitely minute volumes of the gases, which volumes are not
necessarily of equal magnitude, being, in the case of each gas, inversely
proportional to the square root of the density of the gas."[29]

Graham then described the elegantly-simple 'diffusion tubes' which he had
developed. An open, graduated glass tube, $\frac{1}{2}$" wide, and 6" to 14" long, was
fitted with a dense stucco plug. The plug itself was made from slaked Paris
plaster compressed to a thickness of $\frac{1}{5}$" and dried either by heating it to $200\overset{\circ}{F}$
for a few hours or by leaving it out for one day in dry air. The tube was then
filled with water by using a syphon, taking care not to wet the stucco. A
damp filter paper was tied around the outside of the tube to keep the external
air saturated with water vapour. Hydrogen, or another gas, was then bubbled
into the diffusion tube and allowed to exchange with the air outside through
the stucco plug. The constant pressure requirement was achieved by altering
the height of the tube during the experiment to keep the inner and outer water
levels the same. Stucco had a low absorbent power and so Graham's experiments
results were very accurate. The diffusive exchange was completed when the
water level in the tube became steady and Graham then calculated the diffusion
volume for a number of different gases. This was the ratio of the volume of
gas which diffused out of the tube to the volume of air which replaced it.

The results were a clear demonstration of the constant-pressure diffusion law

and, indeed, suggested a useful technique for the determination of specific

gravities of gases, a desideratum of the Chemical committee of the newly

formed British Association for the Advancement of Science.[30]

With regard to an explanation of gaseous diffusion, in 1831, Graham

showed considerable reserve and caution. He had encountered a major difficulty

in his experiments, which left him without an adequate theory to explain his

diffusion law, so he wrote, "the law at which we have arrived .... is certainly

not provided for in the corpuscular philosophy of the day, and is altogether so

extraordinary that I may be excused for not speculating further upon its cause,

till its various bearings, and certain collateral subjects, be fully

investigated."[31]

What was this problem? It was the difficulty of measuring the diffusion

volume of hydrogen which was 3.79 from theory. Dry stucco plugs gave him low

values, such as 3.65, moist plugs gave high values, such as 3.85, and loose,

damp plugs were particularly bad giving him readings as high as 4.05. Only

hydrogen appeared to give such unusual, divergent results so Graham wondered

whether hydrogen possessed some special physical property. This led him to

examine the flow of hydrogen, air, and other gases through a $\frac{1}{2}$" stucco plug

into a small evacuated bell jar. He measured the times taken for the different

gases to reduce the height of an attached mercury gauge from 29" to 27". His

results were most surprising. The velocity of air was almost the same as that

of the heavier gas carbon dioxide. Even more remarkable was hydrogen, which

entered a vacuum 2.4 times more quickly than air, whereas in diffusion

experiments hydrogen passed 3.8 times more quickly than air, or as Graham put

it "a certain quantity of hydrogen passed through the same porous plug, by the

pressure of the atmosphere, into a vacuum in fifteen minutes; by spontaneous

diffusion into air in sixty minutes; or the velocity of diffusion was one-

fourth the velocity of mechanical pressure."[32]

These experiments were clearly quite incompatible with Dalton's first

theory of mixed gases according to which the times of passage of gases by
diffusion into either air or a vacuum should be more or less the same. In a
supplementary note added in 1832 Graham hit on the main reason for this
discrepancy; it was that frictional resistance was involved whenever gases
moved through stucco under pressure.[33]

What was Graham's view of diffusion in 1831? Any answer must be
conjectural as precise documentary evidence is unavailable. However, it is
to be noted that,in both 1830 and 1831, Graham emphasised that diffusion took
place between 'ultimate particles'. Also, in 1831, he wrote concerning
evaporation that "the powerful disposition of the particles of different
gaseous bodies to exchange positions may as effectually induce the first
separation of vapour from the surface of the liquid, as a vacuum would do.
Once elevated the vapour will be propagated to any distance by exchanging
positions with a train of particles of air, according to the law of
diffusion."[34]

To Graham, diffusion illustrated in a very striking way, as he expressed
it, the inherent activity of matter.[35] He developed a vision of diffusion
based on motion. However, it was not the disordered,rectilinear motion of the
later kinetic theory. In 1826, he had argued that caloric particles were
never at rest, even when attached to matter.[36] Rather, caloric particles
vibrated or rotated about a solid atomic nucleus. Therefore when two atoms
collided they would be pushed or thrown apart by the repulsive force exerted
between neighbouring caloric atmospheres. But this was not all, for Graham
emphasised that diffusion was not an accidental process since it was governed
by a fixed law. Therefore the repulsive nature of caloric atmospheres
surrounding atoms was quite insufficient to explain the diffusive motion of
gas particles. The nuclear cores of atoms themselves must be in motion.
Michael Swords in his 1973 Ph.D. thesis has suggested that Graham had probably
worked out an explanation of diffusion by 1831[37] although Graham only revealed
the details of his view in his published speculations of 1863. [38] This seems

to be a reasonable view. I would suggest further that the crisis of 1831,
when no adequate theory was available for diffusion, would provide an
appropriate time for the formulation of such ideas which, interestingly, were
to prove anachronistic to Graham's Royal Society referees, Williamson and
Stokes, in 1863. [39]

Graham believed that the ultimate atoms of elements were identical in
size and weight when at rest. In order to distinguish between the atoms of
different elements he further stipulated that they possessed an unalterable,
and quantitatively-different, motion. This characteristic motion swept out a
constant volume for each atom and hence determined its specific gravity which
was inversely proportional to the volume. Faster-moving nuclear cores gave a
greater velocity of diffusion and so the speed of diffusion of gases was
ultimately determined by atomic motion.

It is possible that the cautious Graham deliberately delayed publishing
his views on diffusion because he was mindful of the fate of his own Glasgow
teacher, Thomas Thomson, whose uncritical espousal of Prout's hypothesis had
provoked harsh criticism from Berzelius, "the acknowledged head of our Science",
as Graham described him. [40]

The response to Graham's researches on diffusion was largely favourable.
In 1833 James David Forbes recommended him for the Keith Prize of the Royal
Society of Edinburgh for the best paper presented to that Society in the past
two years. [41] Berzelius reviewed the German translation of Graham's paper and
noted that there was good agreement between the experimental results and theory
but he cavilled over Graham's quoted accuracy of 1 part in 10,000 for diffusion
volumes, arguing that it should be 1 part in 100. [42] More seriously, in 1857,
Bunsen failed in his attempt to reproduce Graham's diffusion experiments.
This was because he used a stucco plug which was both too thick at 2", instead
of $\frac{1}{5}$", and overheated, which destroyed the essential fine porosity of the
material. [43] Therefore the figures from his diffusion experiments led him
wrongly to question the diffusion law itself. As E.A. Mason has pointed out

172

this may be the reason why Maxwell did not use Graham's stucco-diffusion results in 1860 as fundamental experimental evidence in support of his dynamical or kinetic theory of gases.[44]

Undoubtedly, the law of the diffusion of gases first made the 25-year old Graham well known particularly to Chemists and Physiologists. As Harry Rainy, the Glasgow Physician wrote, concerning Graham in 1837:- "his papers on the law of diffusion of gases .... may fairly rank among the most important contributions which have been made in Chemical Science during the last twenty years."[45]

References

1. T.Graham, "Elements of Chemistry", Baillière, London (1st Edition Part 1, Nov.1837) p.70.

2. J.Priestley, "Experiments and Observations on Different Kinds of Air", J.Johnson, London, (1777), Vol.3, p.305.

3. J.Priestley, Phil.Trans., 1783, 73, 398.

4. J.Priestley, "Experiments and Observations Relating to the Various Branches of Natural Philosophy", J.Johnson, London, (1786), Vol.3,p.65.

5. J.Priestley, Trans.Amer.Phil.Soc., 1802, 5, p.17.

6. J.Dalton, Manchester Memoirs., 1802, 5, 535.

7. I.Newton, "Mathematical Principles of Natural Philosophy", trans. A.Motte and revised by F.Cajori, Univ.of California, Berkeley (1934) p.300.

8. J.Dalton, Manchester Memoirs.,1805, 2 Ser.1 , 259.

9. C.L.Berthollet, "Essai de Statique Chimique", Didot, Paris, (1803), Vol. 1, pp. 483-499.

10. T.Thomson, "A System of Chemistry", Bell and Bradfute, Edinburgh (1st edn., 1802) Vol.3, 270-271 and (2nd.edn.,1804) Vol.3, 314-317.

11. H.E.Roscoe and A.Harden, "A New View of Dalton's Atomic Theory", Macmillan, London (1896) p.134. (letter from Dalton to Hope 8.11.1803).

12. ibid. (letter from Hope to Dalton 29.3.1804).

13. J.Dalton, Phil.Mag., 1804, 19, 80.

14. T.Thomson, "A System of Chemistry", Bell and Bradfute, Edinburgh, (3rd edn.,1807) Vol. 3, p.440, and J.Murray, "A System of Chemistry", Longman, Edinburgh (1809) Vol. 2, 48-53 and Note E.

15. J.Dalton, "A New System of Chemical Philosophy", Bickerstaff, Manchester (1808) Vol.1, p.188.

16. ibid. p.190.

17. J.Dalton, "A New System of Chemical Philosophy", George Wilson, Manchester (1810) Vol. 2, p.560.

18. T.Thomson, "A System of Chemistry", Bell and Bradfute, Edinburgh (4th edn.,1810) Vol. 3, 461-462.

19. H.E.Roscoe and A.Harden, Reference (11) pp.146-150 (letter from Thomson to Dalton 13.11.1809) and C.L.Berthollet, Mémoires d'Arcueil, Paris, (1809) Vol. 2, pp. 463-470.

20. J.Dalton, Phil.Trans., 1826, 116, 174.

21. R.A.Smith, "The Life and Works of Thomas Graham", John Smith, Glasgow, (1884), p.64.

22. T.Graham, "Chemical and Physical Researches", Constable, Edinburgh (1876) pp.28-35.

23. ibid. p.30.

24. W.Henry, "The Elements of Experimental Chemistry", Baldwin and Cradock, London (11th edn., 1829) Vol.2, p.718.

25. T.Graham, "On the Tendency of Air and Different Gases to Mutual Penetration", William Lang, Glasgow (1830) 18 pages.

26. T.Graham, Reference (22), p.44 and p.227.

27. J.Priestley, Reference (5), pp. 19-20.

28. J.W.Döbereiner, Ann.de Chim., 1823, 2 Ser. 24, pp. 332-334.

29. T.Graham, Reference (22) p.44.

30. British Association for the Advancement of Science Report, 1831, 1, p.53.

31. T.Graham, Reference (22), p.69.

32. ibid. pp. 69-70 (supplementary observations dated 7.9.1832).

34. ibid. p.68.

35. T.Graham, Wellcome MSS. 2579, Notes and Drafts for his lectures at
     University College, London. See lecture dated 24.10.1838.

36. T.Graham, Ann.Phil., 1826, 12, pp. 260-262.

37. M.D.Swords, Ph.D. Thesis, "The Chemical Philosophy of Thomas Graham",
     (1973) Case Western Reserve University, pp. 84-87.

38. T.Graham, Reference (22) pp. 299-302.

39. M.Stanley, Ph.D. Thesis, "The Chemical Works of Thomas Graham", (1979)
     Open University, Chapter 7.

40. MS Letter dated 22.2.1837 from Graham to Berzelius, Royal Academy of
     Sciences, Stockholm.

41. MS.Letter dated 6.4.1837 from J.D. Forbes to Graham, Forbes Collection,
     St. Andrews University.

42. J.J.Berzelius, Jahres-Bericht, Tubingen, 1835, 14 pp. 80-84.

43. R.W.Bunsen, "Gasometry: Comprising the Leading Properties of Gases",
     translated by H.E.Roscoe, Walton and Maberly, London (1857) p.203, and
     MS letter dated 9.1.1862 from Graham to C.F.Schonbein, letter 483,
     Universitäts Bibliothek, Basel.

44. E.A.Mason, Phil. Journal, 1970, 7, pp. 99-115.

45. T.Graham, "Testimonials in Favour of Thomas Graham .... Candidate for the
     Vacant Chair Of Chemistry in University College, London." G.Brookman,
     Glasgow (1837) p.42.

The Discovery of Gas–Liquid Chromatography: A Personal Recollecton

By A. T. James

Formerly of UNILEVER PLC, SHARNBROOK, BEDS., UK

I was honoured to be invited to speak to this conference, but
somewhat puzzled, particularly when over the telephone it was
suggested that I talk about the Development of Gas Chromatography.
However, as the conversation proceeded, it became a little clearer
to me: it was my function to provide some little light relief in
a few days of rather hard physical chemistry. So I would like to
try to take you back to the period in which the chromatographic
techniques were being evolved, how they were picked up and how
they were prosecuted. I hope you will find it mildly amusing.

Let me take you back then to before the last war, when the only
separation techniques available to the chemist, the natural
products chemist, or the biochemist, were crystallization (which
of course was only effective if you had sufficient material for
repeated equilibrations), selective extraction, selective
precipitation, or fraction distillation, which even with the most
elegant of the later spinning band vacuum total reflux
distillation columns needed at least five grams of material.
Chromatography is a technique which had been forgotten after
Tswett's initial work, and had been virtually re-invented by Kuhn
and Lederer. This was the liquid-solid system, where the lack of
suitable high-sensitivity detectors often forced the
investigators to heavy column loads, and consequently the
non-linear adsorption isotherms produced distorted zones which
frequently overlapped.

In the hands of Tiselius the general understanding of frontal
displacement and elution analysis was brought forward. Claesson
worked on gas-solid chromatography in the nineteen forties, as
later did C.S.G Phillips at Oxford, but again the technique was

175

limited in application. Chemists were in a slightly more
favourable position than biochemists in so far that, if they had
difficulties purifying their material, they could at least make
more. Biochemists normally had only milligrams of material to
deal with. For these people the older separation techniques were
enormously limiting.

Now in 1941, that changed.

Figure 1 shows the title page of the paper published in 1941 by
Martin and Synge from the Wool Industries Research Association.
Martin and Synge had got together originally at Cambridge.
(Cambridge - I may point out - did not think a great deal of
A.J.P. Martin - he didn't get a terribly good degree.)

Synge was also at Cambridge in the Biochemistry Department.
Martin had done his Ph.D. there with Sir Charles Martin - no
relative - and in fact was the first man to isolate Vitamin E,
but, as was usual with AJP, he never published the work. Martin
was therefore very early on interested in separation techniques of
biologically derived molecules.

Martin had created the very first continuous liquid-liquid
distribution apparatus for the Vitamin E purification. He and
Synge got together to apply liquid-liquid distribution to the
separation of the amino-acids. At that time the great hope was
that, by quantitatively determining amino-acids, all the secrets
of protein structure and function would be revealed (this was
later shown, to say the least, to be somewhat of a simplification).
The modified apparatus they designed leaked chloroform at every
joint so that the man who was sitting with the apparatus was
infernally bad-tempered at the time of take-over so they always
quarrelled then.

The apparatus did not work as well as hoped because of carry over
of both phases due to lack of equilibration in the settling tubes.

Both then moved to the Wool Industries Research Association where

Biochem. J. 35 [1941] 1358

151. A NEW FORM OF CHROMATOGRAM EMPLOYING TWO LIQUID PHASES

1. A THEORY OF CHROMATOGRAPHY

2. APPLICATION TO THE MICRO-DETERMINATION OF THE HIGHER MONOAMINO-ACIDS IN PROTEINS

By A. J. P. MARTIN and R. L. M. SYNGE

From the Wool Industries Research Association, Torridon, Headingley, Leeds

(Received 19 November 1941)

THE MOBILE PHASE NEED NOT BE A LIQUID BUT MAY BE A VAPOUR.

We show below that the efficiency of contact between the phases (theoretical plates per unit length of column) is far greater in the chromatogram than in ordinary distillation or extraction columns. Very refined separations of volatile substances should therefore be possible in a column in which permanent gas is made to flow over gel impregnated with a non-volatile solvent in which the substances to be separated approximately obey Raoult's law. When differences of volatility are too small to permit of ready separation by these means, advantage may be taken in some cases of deviation from Raoult's law, as in azeotropic distillation.

Figure 1   The opening page of the first paper (1) on the liquid-liquid chromatogram which also suggested the gas-liquid chromatogram.

Martin conceived with Synge the idea of holding one liquid phase on
the column packing (in this case silica gel) and moving the lesser
polar solvent through. With this they were able to quantitatively
separate for the first time the aliphatic amino-acids as their
N-acyl derivatives. This first liquid-liquid chromatograph that
really opened up separation technology was published in 1941 (1).

This paper developed the first theoretical plate treatment of a
chromatograph, and showed the reasons for the high efficiencies
obtainable. Martin understood extremely well the physical
chemistry behind the whole system. However the paper also
contained the phrase that "the mobile phase need not be a liquid
but may be a vapour. By means of this, refined separations may
be carried out". This clear indication of a new technique was in
the paper in 1941. Now some of you may suffer from the illusion
that all you have to do is to point a scientist in the right
direction, give him a good idea and he will take it up. Clearly
this was not the case; no one heeded the suggestion.

Now comes a series of coincidences before Martin and I started work
together.

Now, how did we get together? In 1949 I had moved from Bedford
College for Women, where I was working on antimalarials, to the
Lister Institute for Preventive Medicine so that I could work with
R.L.M. Synge on the structure of the cyclic peptide Gramicidin. At
that time scientists were treated rather differently from now as I
shall show. I had also applied to the National Institute for
Medical Research, then at Hampstead, for a position as a
biochemist, and had been offered a job to work in Neuberger's
division of biochemistry at the then quite princely sum of £550 a
year. Since I was attracted to Synge's work on Gramicidin, I
applied for a job at the Lister Institute. I was of course
interviewed by Synge and by the Director of the Institute and I was
offered the job. On enquiring as to what the salary was, I was
told it was told it was £450 a year. Slightly taken aback by this,
I foolishly said to the Director: "Oh, Sir Charles Harrington's
just offered me £550 a year". "Oh", said the Director, "we'll soon

fix that", and in front of me he picked up the phone and harangued Harrington for daring to offer me £100 more than he could afford, so Harrington dropped his by £100: a different world. Anyway, I decided I preferred to go and work with Synge and learn the new chromatographic techniques. Synge left after a year or so and moved to the Rowett Institute in Scotland. Then A.J.P. Martin arrived, to be parked at the Lister Institute while the laboratories at Mill Hill, which had been occupied by the WAAFs during the war, were converted into laboratories and Martin's suite of laboratories was completed. At that time I was working on (as a side issue) attempting a new separation of the amino-sugars which W.T.J. Morgan had asked me to do. The two amino-sugars of mucopolysaccharides, galactosamine and glucosamine, were separated then by paper chromatography (also invented by A.J.P. Martin), but this was laborious and made quantitative only with difficulty. Hence Morgan's request for a better method.

I then developed a new column method using borate complexing with the cis-hydroxyls to separate dinitrophenyl derivatives of the amino-sugars at pH 8. Because of the alkaline conditions needed for the complexing, the column had to be run in a cold room otherwise the coloured bands disappeared. The separation was better than on a paper chromotogram. To stand by a chromatogram in a cold room, at around about $0^{\circ}C$, for a few hours to collect the bands was very uncomfortable. Martin and I gravitated together in the laboratory workshop and he proceeded to invent for me an automatic fraction collector. Such things at that time did not exist. We devised two or three simple mechanical systems operated by a syphon, at the end of a centrally pivoted rotating balance arm so that, as the syphon filled to a constant volume, it dropped, operated a mechanical wire gate - and moved onto the next tube. The syphon then discharged, the arm lifted on to the other side of the gate and waited then for the subsequent filling of the syphon to trip it and move it on again.

This work we published together (2).

Martin then moved to Mill Hill and invited me to join him, as we

got on so well. (Martin I may say was an absolute angel to work
with, a brilliant man, the most brilliant I have ever come across,
an extraordinarily kindly and helpful person, and I enjoyed
enormously the number of years I spent with him.) At Mill Hill the
task given to me was to try and put fractional crystallization on a
column basis. We therefore devised a system of 4 mm columns packed
with Celite, 4 feet long, along which was moved a temperature
gradient cooled at one end with solid CO_2 and heated at the other
end with an electric coil. The mixture to be resolved was put at
the end of a column and solvent was passed slowly along. We were
of course attempting to invent zone refining. I worked on this for
some months and was getting nowhere at all. The technique just
would not work, presumably because of the very slow attainment of
equilibrium in such a constantly filtered system. After three
months' work, the apparatus did less well than by use of
crystallization in the normal way. I thus became somewhat
depressed and this became rather obvious to Martin. Out of
kindness to me he suggested: "Why don't we have a go at the gas
chromatogram?" That cheered me up quite a bit because it was
obvious that we were not going to get anywhere with the continuous
fractional crystallization system. We started with a 4 foot long
4 mm I.D. glass tube, packed with Celite, on which we put a liquid
paraffin stationary phase. We chose, as model substances, to
attempt to separate the volatile fatty acids. We did this simply
because George Popjak, who had the next laboratory to us, was
working on the biosynthesis (in the rabbit mammary gland) of the
short-chain and medium-chain fatty acids. The separation technique
he normally used was the reverse-phase liquid chromatogram (also
invented by A.J.P. Martin), involving the titration of many
hundreds of fractions. Although the resolving power was high, it
took an inordinate amount of time and nearly drove the assistant
staff mad when they had to sit doing the multiple titrations day in
and day out.

The short and medium-chain fatty acids were volatile and they
seemed to be good model substances. Martin, with his usual elegant
simplicity, set up the column to run at room temperature because
that should be adequate for the acetic, propionic and butyric

acids, with nitrogen gas passed through as mobile phase. The end of the tube was drawn round so it passed into a test tube in which was placed indicator in aqueous solution. We simply counted the number of drops from a conical flask containing 0.1N alkali - sufficient to maintain indicator colour. The number of drops was plotted against time, which gave the integral of the normal chromatographic peaks. Much to our surprise, we did not get any separation. In front of the zone was a beard and that continued through to higher concentrations when it immediately overlapped with the beard from the second component, and so on. Very imperfect separation - no clear separation at all. We were totally baffled by this because the gas chromatogram was so obvious after the liquid chromatogram that it should have worked. At first we thought there must have been some very odd adsorption isotherms on the Celite, so we tried in desperation everything we could lay our hands on, even powdered coal. We also tried a capillary column - a simple glass tube with a liquid film on the inside and - to speed up the rate of equilibration - arranged a helical spring inside, which was rotated, through a gland, to stir the film all the time. Still we found the same zone shape - a continuous overlap. This went on for something like two months and we didn't understand these negative results at all. Before finally giving up - perhaps there was something odd about the volatile fatty acids - we thought perhaps we'd try volatile bases, the methylamines.

Figure 2 shows the titration result, plotting drop number against time to maintain neutrality (see Ref. 4). Thus, for the first time, we got the integral of what looked like quite reasonable chromatographic peaks. This was the first really effective gas chromatogram (my code was column 29a, so I must have run at least twenty-eight others beforehand). We thought we might still be getting slight absorption so we treated the Celite with alkali and improved the zone shape to that expected from a gas-liquid linear distribution isotherm.

That then gave us the first real way of showing that the gas chromatogram was perfectly feasible and that refined separations of very small amounts of material were achievable.

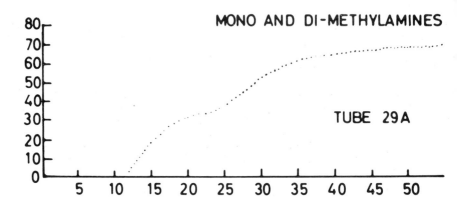

<u>Figure 2</u>   One of the very first successful gas-liquid
chromatograms showing the separation of mono- and
di-methylamine, plotting number of drops of dilute
acid required to maintain indicator colour against
time.

<u>Figure 3</u>   The automatically recording microburette using a
photocell to scan the indicator solution and so
control electronically the burette.

Now, of course, this was very laborious - you had to sit at the
bench with the simple little drop titrator and with a watch.
Martin used to have the watch, I used to have the drop titrator.
But of course, if somebody came in to talk to you (which at Mill
Hill was a pretty frequent occurrence) you turned round and lost
the sequence, and the separation had to be repeated.

We clearly needed a much simpler titrator, so Martin proceeded to
devise a very elegant burette in which one rotated a handle fixed
to a lead screw which forced a stainless steel rod into the burette
body itself through a simple polyethylene gland. By turning the
handle a few times to get back to neutrality, one could read off
the volume delivered on a linear scale. That was a great
advantage; we could do many more experiments and study the
separations more easily. Nevertheless, it was still laborious:
you had to sit at it and watch the indicator solution continuously.
The next stage was to make an automatic recording burette, and this
we did.

Figure 3 shows the electronically controlled microburette based on
the earlier microburette. A ball point pen fixed to the lead screw
plotted automatically, on a drum rotated by an electric clock
movement, the volume of titrant delivered.

In Figure 4 is shown a schematic diagram of the whole apparatus.
The column was held in a vapour jacket with a simple pressure seal
at the junction with the titration cell.

With the vapour-jacketed system we could vary the temperature of
the column by changing the refluxing solvent. In those days one
could not run along to a laboratory supplier and get electrically
heated, thermostatically controlled devices - you either made them
yourself or you did without. We used an air condenser so that we
didn't have cold condensate dripping back on the column to give a
cold spot. With this apparatus we were able to study a wide range
of volatile substances capable of being quantitatively detected by
titration.

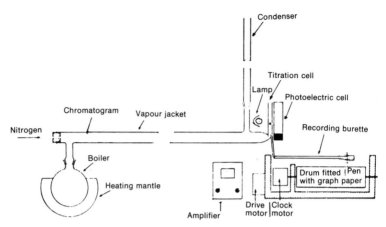

– *Schematic diagram of apparatus for gas-liquid chromatography, using*
automatic *titration to detect volatile acids or bases emerging from the column.*

Figure 4   Schematic diagram of the first automatically
                  recording gas-liquid chromatogram using vapour
                  heating of the columns.

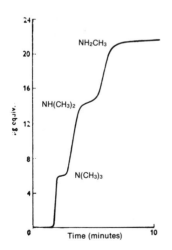

– *Separation of mono-, di-, and tri-methyl-*
amines as recorded by the automatic burette. Column tempera-
ture 78·6 °C; stationary phase glycerol.

Figure 5   Rapid separation of mono-, di- and tri-methylamines,
                  using the apparatus in Figure 4.

The device is now in the Science Museum, on the first floor. I
don't recommend putting one's original apparatus in the Science
Museum - it's like erecting your own tombstone!

Figure 5 shows an example of a rapid separation of trimethylamine,
dimethylamine, and methylamine, giving complete separation between
zones.

Figure 6 illustrates the effect we were getting with the fatty
acids. In Curve A the zones were bearded at the front with the
high concentration part of the zone overlapped with the beard from
the subsequent peak. We sat and discussed and argued over this for
weeks as to why the fatty acids were behaving anomalously. Then
the penny dropped, obvious now. Under the conditions we were using
the short-chain fatty acids were dimerizing at the high-
concentration part of the zone so the partial pressure became
proportional to the square of the concentration in the stationary
phase. At the low-concentration part of the zone, the distribution
coefficient was much more in favour of the moving phase. Low
concentrations were travelling faster than the higher
concentrations. The high concentration part of the zone had a
sharp back. The obvious way to deal with this phenomenon was to
add a long-chain fatty acid to the stationary phase, (which would
not effect the titration cell). This swamped out the effect and
allowed clear separation between the zones. The paper describing
the separation of the methylamines (4) was actually published after
the fatty acid paper (5) and not in historical sequence.

This work demonstrated the refined separation now possible for
volatile compounds. In the first paper, the theory of the gas-
liquid chromatogram was developed covering the compressibility of
the moving phase and the theoretical plate concept was also
covered.

However the device we had constructed was limited to titratable
substances. I described the technique at a Biochemical Society
meeting at Mill Hill in October 1951. The first full paper was
published in 1952 in the Biochemical Journal (3). I went on to

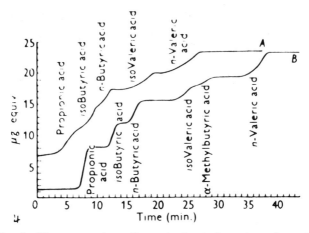

Fig. 6. The separation of propionic, *iso*butyric, *n*-butyric, commercial '*iso*valeric' and *n*-valeric acids. *A*, column length, 4 ft.; liquid phase, DC'550 silicone; nitrogen pressure, 30 cm. Hg.; flow rate 23 ml./min.; column temp., 100 . Incomplete resolution of all the acids. *B*, column length, 4 ft.; liquid phase, stearic acid (10% w/w) in DC'550 silicone; nitrogen pressure, 46 cm. Hg; flow rate, 45 ml./min.; column temp., 100 . The '*iso*valeric' band has been almost completely resolved into two components.

Figure 6   Separation of short chain fatty acids showing the
                  effect of incorporation of stearic acid in the
                  stationary phase in suppressing dimerisation
                  effects, Curve A; no stearic acid and distorted
                  zones, Curve B. Effect of stearic acid in the
                  stationary phase undistorted zones.

make an intensive study of chromatographic behaviour of a whole range of amines and Figure 7 shows that, by comparing the relative retention volumes in two stationary phases - a non-polar one, liquid paraffin, and a polar, one could differentiate between amines of different type - e.g primary, secondary and tertiary. For a time, this procedure of using different stationary phases was called the "poor man's mass spectrometer" since, by careful study of chromatographic behaviour of an unknown compound, one could get a great deal of indication of its structure (5).

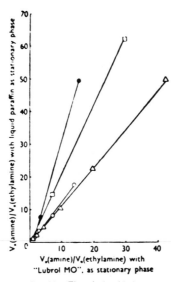

Fig 15. The relationship between (retention volume of amine/retention volume of ethylamine) for columns having stationary liquid phases of (a) liquid paraffin and (b) "Lubrol MO," showing the relative speeding up of secondary and tertiary amines on changing from a non-hydrogen bonding solvent to one allowing hydrogen-bonding

— △ — Primary straight-chain amines
— ○ — Primary amines with an iso-configuration
— □ — Secondary   straight-chain amines
— ● — Tertiary straight-chain amines

Figure 7   Differentiation between primary, secondary and tertiary aliphatic amines by comparison of chromatographic behaviour on different stationary phases.

We now tried to get an even high resolving power and constructed a
column 11 feet long, which enabled us to separate the whole range
of acids from C_2 to C_5 including isomers   (Figure 8). In this
diagram the integral curves have been graphically differentiated to
give more characteristic peaks. A comparison is made of the
resolving power of 4 and 11 ft columns. Thus, the first papers
had demonstrated the understanding of the effects in solutions and
how they could affect chromatographic behaviour. We showed that
a series of straight chain compounds had a constant increment in
retention volume for each additional CH_2.

Having developed a new separation technique widely applicable, we
were still limited by the nature of the gas detector. A general

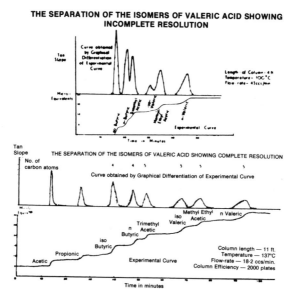

Figure 8   Comparison of 4 feet and 11 feet columns in the
                separation of isomeric fatty acids.

detector did exist, the catharometer, but Martin disliked the
instrument because it was flow sensitive and needed to be
calibrated for every substance to give quantitative results.
Nevertheless it was well known and was of quite reasonable
sensitivity. We proceeded to construct a large column capable of
handling grams of material. It was 20 feet long and an inch in
diameter, which took me about two days to pack. We were able to
separate five gram quantities of the C_5 and C_6 aliphatic
hydrocarbons very readily at room temperature, much more
efficiently than could be done on the very best of total reflux
fractionating columns. For this Martin devised a different
catharometer - one in which the flow was at right angles to the
axis of the wire which made it inherently much less flow-sensitive.

The small experimental fluctuations in flow rate didn't affect the
qualitative response. It was a much less noisy device than the
conventional catharometer. Nevertheless, it was pefectly good for
detecting the presence of a zone and for allowing us to collect it
by condensing it out from the gas stream in a cooled trap. We did
not publish this work, like much else of our experimental work.

Martin became fascinated by the possibilities of condensing out on
a balance pan cooled to a low temperature the material evolved from
the column because this would give a quantitative mass
determination straightaway. Such a device however would require a
microbalance of quite exceptionally high sensitivity and of a
quality which had never been worked on or produced before. There
was nothing one could buy of comparable sensitivity. Martin
proceeded to design a single-pan quartz torsion balance. This
meant using very thin quartz rods both for the arm and, after
thinning, as thin torsion threads. Such manipulation was so touchy
that one could do it only for about half an hour in the morning
because one's hands actually showed enough tremor to break the
rods. The whole thing was so fragile, the beam being made of 0.5
mm quartz rod, that you only had to sneeze - as I did one day - and
the whole thing shattered.

Nevertheless, Martin persisted with this, but I became more and
more convinced that it was never actually going to be feasible.
Therefore I got him to go back to a device he had thought of many

years ago. Claesson, who had worked on gas-solid chromatography
had used a gas density balance as detector. It consisted of a
tubular device of an enormous dead volume, about 200 mm^3 - enough
to spread any zone - and a rather low sensitivity - but it did
work.

Martin went back and thought again for some weeks about his concept
of a gas density balance.

Figure 9 shows a schematic diagram of the gas density balance.
This was a beautiful and most elegant solution to a problem
(typical of Martin). The gas density balance was a three-
dimensional set of channels giving two parallel moving columns of
gas, one from a reference column with the same stationary phase, the
same length and the same pressure drop, and the other from the
chromatograph column. As soon as there was a density difference
between the effluents from the columns due to the presence of a
solute in the gas phase, this caused a movement of gas from one
stream to another to restore equilibrium through the cross channel

GAS DENSITY BALANCE.

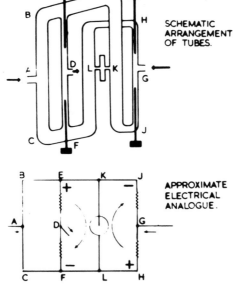

Figure 9   Schematic diagram of the gas-density balance.

L^1-L. This flow was detected very simply by a small electrically
heated wire loop detector which gave a vertical stream of convected
gas rising to a double thermocouple copper-constantan suspended
above the loop and clamped into position. As soon as there was any
gas flow, the hot convected stream was deflected, one thermocouple
became hotter than the other and produced an electrical signal.
This was amplified and fed to a recording galvanometer. This
device could be calibrated absolutely in terms of the difference in
molecular weights of solute and carrier gas. Furthermore, when
adjusted correctly, it was flow independent and more sensitive than
the catharometer.

The gas-density balance was fitted inside the same vapour jacket as
the columns and the whole arrangement is shown schematically in
Figure 10 and in a photograph in Figure 11.

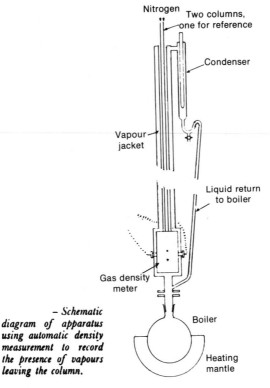

*– Schematic
diagram of apparatus
using automatic density
measurement to record
the presence of vapours
leaving the column.*

Figure 10   Schematic diagram of the columns, heater and gas
density balance.

Figure 11   Photograph of the actual apparatus.

Martin's approach was to think up an idea, which he wished to resolve experimentally. He and I would discuss it for some days; at the end of about three or four days I would say: "Let's do the experiment", to which the answer always was: "No, no, no – sit down and talk about it some more". Over a period of perhaps two or three weeks we worked through mentally many, many experiments until in the end we did "the" experiment which defined it. The gas density balance exemplified this approach of Martin's. He thought it through, the whole thing, again and again, until having spent many weeks thinking he constructed a simple glass model. Now while I was quite a reasonable glass blower, Martin was an expert. Whenever I made any apparatus it worked, even if it looked as if it had been hewn out of wood – he would just take it over and heat it in the flame and revolve it a few times and then the joints looked perfect. The glass model was meant merely to demonstrate the principle, and the principle worked. The next device was a half-scale model constructed out of a block of brass. The third one was finally constructed out of a block of copper so that with the high conductivity of the copper no parasitic effects due to thermal differences in the gas streams would occur.

We then went on to show that this was now a general purpose detector which could detect any vapour in any gas stream and its response was easily calibrated for all substances. The paper (6) was a devil to write, especially the diagrams, since the device was bored from a three-dimensional block. Explaining this device to visitors when it was a visually impenetrable block of copper was somewhat difficult, so we made two or three transparent perspex models which then travelled all over the place, particularly to industrial companies who became interested.

We now could expand out into a range of other substances.

Whilst we were attempting the microbalance approach, we were visited by N.H. Ray of ICI who had seen our early work. Martin suggested to him that he make an infra-red detector as a very useful extension to existing detectors. Ray went back but used a catharometer and published a whole series of very good separations – in fact the first, I think, of alcohols and aliphatic halides. Martin and I now thought we would look at hydrocarbons because we

were very interested in chromatographic behaviour and the rules
governing it. Consequently we tried to separate the components of
40/60, 60/80, and 80/100 petrol. In Figure 12 is shown the
resolution we got; peak 5 in curve A is isopentane, 6 is normal
pentane, 8 and 9 are isomers of hexane. That was all very well,
but we couldn't identify all the peaks we were getting in 80/100,
or even in 60/80 petrol. Now we knew that during the war there was
an enormous collaborative scheme among organic chemists from the
United States and Britain, aimed at the synthesis of all isomeric
hydrocarbons up to and beyond C12. We now discovered that this
collection of pure hydrocarbons was reposing in BP. In order to
relate chromatographic behaviour to paraffin hydrocarbon structure,
we asked BP if we could have samples of all the C_5 to C_8
hydrocarbons. They wrote back very politely and enquired as to
what we proposed to do with them. We replied by sending them these
three separations, identifying some peaks and explaining that these
were the first gas chromatographic analyses of paraffin
hydrocarbons. Two days later, the Director of Research of BP and
Denis Desty arrived in our laboratory, and from then on BP never
looked back. They soon scrapped their spinning band distillation
columns and replaced them by gas liquid chromatography under the
leadership of Desty who developed this field extensively.

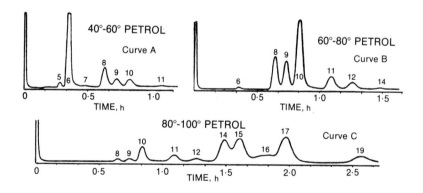

Figure 12   Early separations of the components of 40-60°,
                   60-80°, and 80-100° petrol using the gas-density
                   balance as detector.

(As an interesting aside, the American who produced the world's
best spinning band distillation columns, whom Martin knew, visited
us in the early days of the gas chromatogram, when he virtually
monopolised the world market in high-efficiency fractionating
columns. He looked at the device and said: "Very ingenious, but
it's of no interest to me". But seven years later I don't think he
had a lot of business, the gas chromatogram having taken over so
much from the fractionating column.)

Now that we had a whole set of compounds to work on, I studied
extensively their chromatographic behaviour. We had already shown
that addition of a methylene group to the backbone in fatty acids
(Figure 13) and amines give a constant increment in retention volume.

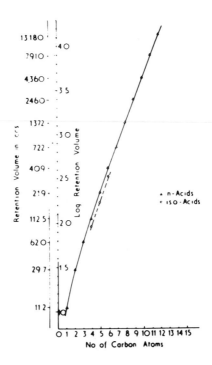

**RELATIONSHIP BETWEEN LOG RETENTION VOLUME
AND NO OF CARBON ATOMS IN THE
VOLATILE FATTY ACIDS.**

Figure 13   Relationship between log retention volume and chain
length for aliphatic fatty acids.

The same was now shown to be true for the hydrocarbons, even in
branched chain and cyclic structures (Figure 14). This was
published in 1956 (7), rather a long time after we completed the
work, as usual. During all this time we had been visited by many
people, Roy Scott, the Unilever groups, Al Zlatkis and many others
who went on to further develop the technique.

Martin had now decided to leave the National Institute and move
into industry because he was interested in the development of
improved instruments. I continued at the National Institute and
concentrated on the long-chain fatty acids, their separation and
quantitative estimation. I was rapidly able to show very highly
refined separations (Figure 15) which could pick up a whole series
of fatty acids previously unknown to occur naturally. I then
switched from gas chromatography to biochemistry, in particular the
biochemistry of the fatty acids. By combining simple chemical
additive and oxidative reactions, I was able to identify the
position of double bonds in the molecule on a micro scale (8).
Biosynthesis pathways were, at that time, most easily studied by
use of radiotracers (especially ^{14}C), so Piper and I developed a
radio gas chromatogram using a proportional counter as detector

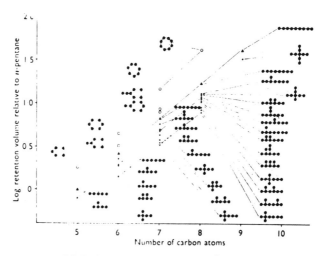

*- Relationship between structure and rate of movement down the column
(relative to n-pentane) of a variety of saturated hydrocarbons. Column temperature
65° C; stationary phase n-octadecane.*

Figure 14   Relationship between log retention volume and
               structure of a range of hydrocarbons.

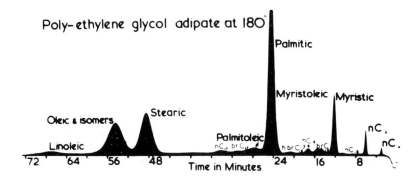

Figure 15   Separation of long chain fatty acids as their methyl esters.

(9). I had at one time eight of these in the lab, which we built ourselves, since none was commercially available. Figure 16 shows the type of record attainable.

I now began a series of collaborations, having worked out these methods of separating and identifying the structure of fatty acids on a micro-scale. One of my earliest collaborators was Jim Lovelock, also at the National Institute for Medical Research but in a different division. Lovelock persuaded me to make a study of the fatty acids in the lipids of the blood of patients with coronary artery disease, because there were many theories but little practical experimental fact. We did a study with some clinicians, and showed indeed that fatty acid neutral lipids were elevated in those patients who had at least one infarction. However, we needed 5 mls of blood to separate out the neutral lipids from the plasma phospholipids. Lovelock suggested a much more sensitive detector based on ionisation properties of the effluent gas from the column.

We rapidly evolved the first argon ionisation detector – an order of magnitude greater in sensitivity than the gas density balance. My major contribution was the central electrode which was actually a tractor spark plug. Lovelock went on to develop even more highly sensitive ionisation detectors of low dead volume and this really

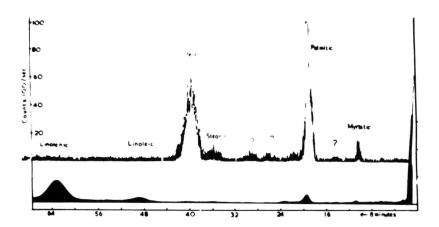

<u>Figure 16</u>   Radiochemical analysis of fatty acids eluted from
                       the gas chromatogram.

opened up the field, especially for capillary GLC developed by
Golay. During the time from 1952 onwards the work was rapidly
taken up elsewhere. Our very happy period, in which for the first
three or four years we knew more about the subject than anybody
else, rapidly disappeared but I was happy to devote myself to lipid
biochemistry and subsequently biotechnology.

Figure 17 shows A.J.P. Martin, Nobel Laureat. This man changed the
face of chemistry, analytical chemistry and biochemistry, perhaps
more than any other. His development of the liquid-liquid
chromatogram, the reversed phase chromatogram, the paper
chromatogram, and the gas-liquid chromatogram, make him the finest
genius that I have every met. We owe him a great deal for his
work.

Figure 17   A.J.P. Martin in about 1954 at Mill Hill.

REFERENCES

1. Martin A.J.P. and Synge R.L.M. (1941). Biochem. J. 35, 1358.

2. James A.T., Martin A.J.P. and Randall S.S. (1951). Biochem. J. 293.

3. James A.T. and Martin A.J.P. (1952). Biochem. J. 50, 679.

4. James A.T., Martin A.J.P. and Smith G.H. (1952). Biochem. J. 52, 238.

5. James A.T. (1952). Biochem. J. 242.

6. Martin A.J.P. and James A.T. (1956). Biochem. J. 63, 138.

7. James A.T. and Martin A.J.P. (1956). J. Applied Chem. 6, 105.

8. James A.T. and Webb J. (1957). Biochem. J. 66, 515.

9. James A.T. and Piper E.A. (1961). J. Chromatography, 5, 265.

The Story of Low Temperature Gas Separation

By C. G. Haselden

DEPARTMENT OF CHEMICAL ENGINEERING, THE UNIVERSITY OF LEEDS, LEEDS LS2 9JT, UK

Introduction

In low temperature gas separation one is fighting the Second Law of Thermodynamics twice over, so nothing comes easily. It took two remarkable characters, Carl Linde and Georges Claude to make it a reality. In fact they were fighting on three fronts because the measure of their success was not just that they separated gas mixtures into their constituents at previously inaccessible temperatures, but they did so profitably. They were engineers rather than scientists.

The possibility of economical low temperature gas separation rests on three related scientific facts. The first is that liquefied gases are, to a very high degree, mutually soluble. The second is that such solutions when in equilibrium with the vapour phase normally display a significant change of composition between the phases. The third, and equally important fact, is that the processes of molecular bombardment at such vapour-liquid interfaces are extremely rapid. Therefore one can have high rates of molecular flux across the interface, in a process such as distillation, without disturbing the equilibrium measurably. This point was recognised after the event.

Gas Liquefaction

The task of persuading gases to go into the liquid phase was undertaken with zeal by Michael Faraday who by 1845, and using a combination of compression and cooling with solid carbon dioxide, had forced all the known gases except oxygen, nitrogen, hydrogen, carbon monoxide and methane to succumb. It took more than thirty years before, simultaneously in December 1877, Louis Cailletet in France and Raoul Pictet in Switzerland demonstrated that oxygen could be condensed. In both cases the evidence rested on mist formation for a brief instant. They could not produce enough liquid to collect. This achievement is credited to Sigmund von Wroblewski and Karel Olszewski in Poland in 1883. Wroblewski also liquefied air two years later, and was the first to observe that it formed a homogeneous solution. Before then it had been assumed that the miscibility of liquid oxygen and nitrogen was small.

From this time competition to liquefy air on a significant scale became intense. Carl Linde in Germany, Tripler in the USA and William Hampson in England, though working quite independently, all arrived at approximately the same solution in 1895. In each case the feed air was compressed to about 200 atmospheres pressure and was water-cooled back to atmospheric temperature, thereby reducing the enthalpy of the air below its starting value (though they did not recognise the fact). The compressed air was passed through a heat exchanger to an expansion valve. The small Joule-Thomson cooling effect was made cumulative by returning the cooled air

201

through the other pass of the exchanger. Hence the valve progressively cooled down until (if the exchanger was good enough) a small fraction of liquid air was produced. Fig.1. shows the Hampson liquefier with its spirally wound heat exchanger. Many such units were sold by BOC for laboratory use. They produced liquid within ten minutes of starting and had a steady state production rate of about 1 litre per hour.

Carl Linde (Fig.2) was born in 1842 and had worked in mechanical

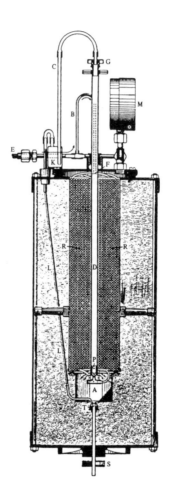

Figure 1.

The Hampson Liquefier for air. Air at 200 bar entered at E and flowed down through the spirally wound heat exchanger R to the expansion valve P. Liquid air accumulated in A and could be withdrawn through valve T, emerging at S.

Figure 2. Carl von Linde

Figure 3. Georges Claude

refrigeration for several years before, in Wiesbaden in 1879, he founded a joint stock company named Gesellschaft für Linde's Eismaschinen.   The company grew rapidly and provided the base on which he developed his low temperature interests.   In 1890 he handed over the chairmanship of the Company to a colleague and went back to Munich as a college professor.   It was in this rôle that he invented the air liquefier, followed shortly afterwards in 1902 by his first oxygen plant.   These achievements earned him an honorary doctorate of Braunschweig College of Technology and elevation to the nobility.

Georges Claude (Fig.3) was a younger man, being born in 1870, and was trained as an electrical engineer.   At age 26 his work involved him with the problem of storing acetylene.   Previously Pictet had tried to store it as liquid in a shop in Paris, with catastrophic results.   Claude noted the efficacy with which a soda syphon trapped carbon dioxide and by analogy decided to experiment with solvents for acetylene.   Thus he hit on the use of acetone – which has been used safely ever since.

It was the potential of burning acetylene in oxygen to generate intense heat which caused Claude's interest in oxygen.   He realised that atmospheric air was the obvious source.   So he studied the work of Carl von Linde but disapproved of the extremely high pressure (200 bar) at which his process operated. He recognised that if the expanding air could be made to do work in an engine, rather than being throttled in a valve, it would be possible to achieve higher degrees of liquefaction with much lower air pressures.

He chose a maximum air pressure of 40 bar and experimented with various forms of expansion engine substituted for the throttle valve in a Hampson type cycle.   As the inlet temperature went down the engine become more and more prone to seizure and in addition the work output greatly diminished. He found that by progressively diluting the lubricating oil with petrol as the temperature went down he could keep the engine running; however the non-ideality of the air under these conditions resulted in little work output or cooling.   Then he found that it was better to precool the air only slightly before engine expansion, and to use the expanded air to cool and liquefy another fraction of air at 40 bar which he then expanded to atmospheric pressure through a valve.   He also found that piston sealing rings made from some grades of oil-free leather (Kangaroo was particularly successful) retained flexibility at very low temperatures and could be used without lubrication.

Thus Claude developed an efficient and reliable liquefier (Fig.4) with a claimed performance of 0.75 litre of liquid air per kWh compared with 0.70 litres for a Linde liquefier (Fig.5) which had a dual pressure expansion and ammonia precooling.

Distillation of Liquid Air

In 1902 Linde first used a distillation column (Fig.6) to produce nearly pure oxygen.   Under steady state conditions a feed air pressure of about 50 bar was required to combat heat inleak and to maintain the inventory of liquid.   The gas leaving the top of the column contained about 7% O_2 and so the recovery of oxygen from the feed air was only about 70%.

By 1910, with the help of Rudolph Wucherer, Linde had developed the double column (Fig.7) with the lower one operating at about 5 bar and the upper one at just over 1 bar.   The development of this device, which is well described by Ruhemann,[1] involved much more than merely raising the

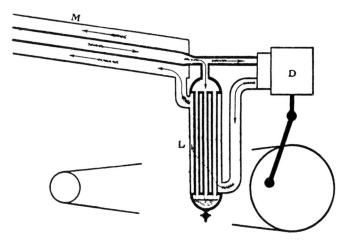

Figure 4.

The Claude Liquefier. The feed air at 40 bar is cooled in exchanger M. The greater part of it is expanded in engine D and used to cool and condense the remainder in the liquefier L, from which it is drawn off as product.

condensing temperature of the nitrogen in the lower column to a value above that of the oxygen boiling at the base of the upper column. It implied also a very elegant appreciation of the reflux and reboil requirements of the two parts of the column, and the way in which, fortuitously, they can be mutually satisfied for the air system. It is not a general solution to all gas separation problems, though the double column has been proposed for one or two other mixtures.

Claude tackled the distillation problem at much the same time and by 1905 had built a plant producing about 100m³/hr of oxygen of 93% purity. His cycle used a reflux condenser (or dephlegmator) in place of the Linde high pressure column. The feed air pressure was set initially at about 40 bar and the greater part expanded in the engine down to about 5 bar to provide cooling to quickly accumulate liquid. The expanded air ascended an array of vertical tubes surrounded by crude liquid oxygen. Partial condensation occurred under refluxing conditions leaving mainly nitrogen gas at the top of the tubes and a rich liquid containing more than 40% oxygen as bottom product. The nitrogen was condensed as it flowed downwards in another array of vertical tubes which were placed around the first set. The resulting liquid nitrogen was then used as reflux for the low pressure column, whilst the rich liquid provided the centre feed. Under steady state conditions, for a plant producing gaseous products, the feed air pressure could be dropped to 18 bar.

This apparatus was simple and thermodynamically attractive but the fractionating action of the dephlegmator was not very great. Hence the nitrogen purity (and the oxygen recovery) was not as high as with the Linde Double Column.

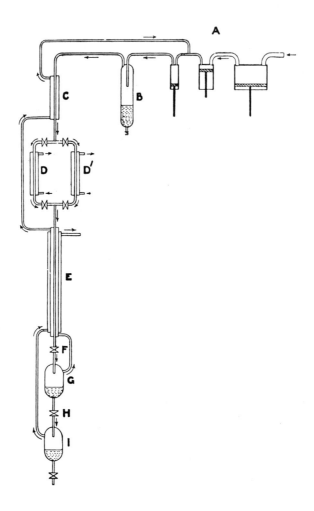

Figure 5

An advanced Linde Liquefier. Air from a 3-stage compressor at 200 bar has most of the moisture removed in B, is precooled in C before going to duplicate ammonia coolers D. After further cooling in E it is expanded to about 25 bar and partially liquefied. The gaseous fraction is returned through the heat exchanger to the final stage of compression whilst the liquid is further expanded to provide the product.

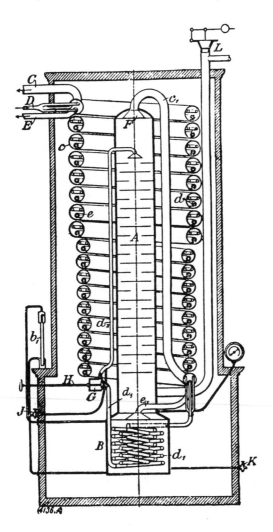

Figure 6.

The Linde Single Column for oxygen production.   Air at 50 bar enters the coiled heat exchanger at D and then is condensed in the coil d, in the reboiler.   It is then expanded through valve G and injected as reflux into the top of the column.   The liquid accumulating in the shell of the reboiler is nearly pure oxygen.   The gaseous oxygen product is collected by e, and leaves through pipe E.   Rectification occurs from plate to plate in column A.

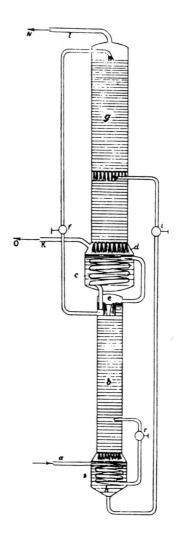

Figure 7.

The Linde Double Column.   The feed air at about 6 bar is condensed in the
coil in the lower reboiler and is fed at an appropriate level in the high
pressure column b.   The rich liquid collecting in the shell of that
reboiler is used as feed to the upper low pressure column g.   Due to the
pressure difference the liquid oxygen collecting in the shell of the upper
reboiler is able to condense the nitrogen from the top of the lower column,
and hence provide reflux through valve f for the upper column.   Oxygen
product is indicated by the arrow labelled 0.

At this time there was litigation concerning the standing of the rectification patents of Linde and Claude and it was found that Claude's ideas were in some measure derived from those of Linde. An agreement was made between the companies which enabled Claude to go ahead manufacturing and operating oxygen plant at the Company 'L'Air Liquide' which he had formed with Paul Delorme in November 1902.

Two further attributes of both Linde and Claude deserve recognition. Firstly both men, with their staffs, displayed outstanding mechanical ingenuity. They needed to develop hardware to do new things at unknown temperatures with untested constructional materials. Secondly they supervised an extraordinary rate of growth of the Companies they founded. For instance the hourly oxygen capacity of Linde's German works at the end of 1903 was $20m^3$. Five years later it had risen to 128 and by 1913 it was $835m^3$. Simultaneously the first UK oxygen plant to Linde design was installed for the newly formed British Oxygen Company at Birmingham in 1906. By 1913 there were six other installations in the UK with a total output of $300m^3$. The first plant installed by Linde in the USA through its new US subsidiary was at Buffalo in 1907 and by 1913 there were plants in seven other centres. There were also plants operating by this date in nine other European countries as well as in Russia, the Argentine, Chile, India, China and Australia. Few new industries can parallel such growth rates.

The Development of Air Separation

The 20's and 30's saw a consolidation of oxygen production on a worldwide basis, the addition of rare gas recovery, the beginnings of liquid oxygen transport (liquefaction being based on the Heylandt cycle), and the emergence of 'tonnage' oxygen. Tonnage impled a production of at least 100 tons per day and was related to new demands in iron and steel making and in fuel conversion. The required changes to the air separation plant were the introduction of regenerators and expansion turbines.

By this time it had been made clear by Helmuth Hausen and others that an air separation plant had to satisfy three different, though interrelated, operating criteria. Firstly the reflux and reboil requirements of the double column required to be met so that products of the required purity could be generated with a reasonable number of plates in each column. Secondly the refrigeration requirements of the plant had to be met. These were of two sorts; there were heat exchanger losses due to the fact that the outgoing products always left at a temperature slightly lower than that of the ingoing air, and there was heat inleak through the insulation. Traditionally Linde had solved this problem by compressing at least part of the feed air to about 200 bar, and preferrably by reinforcing the resulting Joule-Thomson effect by precooling the air with ammonia. The new challenge was to admit all the air at a pressure of not much more than 6 bar and to obtain cooling by expanding part in a turbine (a reciprocating expander would have been far too large for the volume of gas to be handled). This challenge involved not only the building of a suitable turbine but it required very close attention to distillation needs, because the feed air expanded in the turbine, and injected directly into the low pressure column, was not available to generate reflux and reboil. Thirdly the plant needed to provide a cheap and convenient way of separating moisture and carbon dioxide from the feed air, since otherwise they would rapidly cause clogging. Up to that time packed beds of desiccants such as caustic soda had been used to scrub the feed air - at high expense in labour and chemicals. With a low feed air pressure the purification problem was exacerbated.

Hence the pioneering work of Mattias Frankl, begun in 1925, on regenerators was most timely.   He developed a very efficient 'pancake' packing design. It used corrugated aluminium ribbons, and incorporated a lot of heat transfer surface whilst presenting only a small pressure drop.   The alternating flow streams and rapid switching meant that the moisture and carbon dioxide which precipitated in the appropriate zones of the packing did not migrate and could be re-evaporated into the outgoing product streams. The conditions under which the precipitate could be fully re-evaporated were, however, very stringent because at any point in the regenerator the temperature of the returning product stream was always lower than that of the depositing air stream flowing in the reverse direction.   The resulting lower vapour pressure could only be compensated by the fact that the product flows were at lower pressure, and therefore had a larger volume.   This problem is known as 'clean-up' and is greater for carbon dioxide than for moisture.   Of course the oxygen product from regenerator plants was not of high purity because it contained not only small amounts of water vapour and CO_2 but also some air picked up at switch-over.

In early Linde-Frankl plants a few per cent of the feed air was compressed to high pressure and chemically purified so that it could be expanded in an engine or in a throttle valve (after ammonia precooling) to supply the refrigeration needs.   This extra air entered the high pressure column and so was available to help distillation and clean-up.   The avoidance of the high pressure stream required that part of the low pressure feed air be available to enter a turbine at a temperature somewhat below $-100°C$ so that the expanded air, at saturation temperature, could enter the low pressure column.   To do this one had either to withdraw some air at an intermediate position in the regenerators and remove residual CO_2, or one had to rewarm part of the purified cold air by coils buried in the packing at the low temperature end of the regenerators.   This problem was beginning to be addressed when the Second World War intervened.

Turbines

The first expansion turbines used in oxygen plants were designed and built by Linde but they were based on steam practice and were inefficient.   The credit for analysing the special requirements of low temperature turbines goes to Peter Kapitza of the USSR. In fact he was working in Cambridge during part of this time.   Due to the high density of air, and the resulting lower gas velocities, it is best to use a radial flow design thus capitalising on the large centrifugal contribution.   The size of the required turbine proved to be very small.   For an air flowrate of $1000m^3$/hr., measured at NTP, Kapitza's turbine had an outside diameter of only 80mm and ran at 40,000 rpm.   The nozzle blades were curved but for simplicity the blades on the rotor were made straight.   Close attention was paid to bearing design and balance.   As a result, despite the very high rotational speed, his turbine was found reliable and to have an isentropic efficiency approaching 80%.   All later industrial cryogenic turbines were based on this design.

Developments of Air Separation Since the 1950's

Considerable innovation in air separation plant occurred in the early 1950's in the USA based on wartime experience with ingeneous mobile plants. Professor Samuel Collins at M.I.T., supported by others including Howard McMahon of Arthur D. Little Corporation, had developed a very efficient small reciprocating expander and various designs of reversing exchanger. The expander was suitable for relatively low inlet pressures, say 6 bar. The reversing exchangers, in contrast to regenerators, allowed uncontaminated high purity oxygen to be produced.   It also allowed a fraction of the cold and purified feed air (the so-called Lachmann air) to be partly rewarmed in an extra heat exchanger pass to provide the expander feed.   The new reversing exchanger designs were also very compact.

Much was written at this time concerning new tonnage oxygen plant designs but few were built.   However by the middle 1950's large reversing plate-and-fin exchangers, made of aluminium using salt-bath brazing, were being produced by a small number of manufacturers in both the USA and Europe, and these replaced regenerators in most new plants.

The throughputs of plants increased and reached values as high as 1000 tons of oxygen a day.   Almost all the components were fabricated from aluminium and most joints were welded.   These components were built into 'cold-boxes'.   In some cases these were designed as rigid structures into which all the plant components were assembled and pressure tested in the manufacturers workshop.   Then they were transported hundred of miles and finally erected on site.   The slag-wool, which had been the traditional cold-box insulant, was replaced by powdered 'Pearlite' which had a lower thermal conducitivity and could be produced on site and handled pneumatically.

Safety became a matter of great concern from two angles.   Firstly there was an explosion risk due to combustible materials accumulating in the liquid oxygen within the plant.   Secondly there was the risk that enriched air containing 40-50% oxygen would condense within the insulation surrounding the upper part of the low pressure column (which was colder than the dewpoint of the surrounding air) and would mix with any combustible components in the insulation.

The accumulation of hydrocarbons in the liquid oxygen was promoted by the long operating periods of the plants, by their location in the vicinity of refineries and petrochemical sites, and by the use of low pressure purification processes which allowed traces of acetylene and ethylene to slip through.   These hydrocarbons precipitated as finely divided solids in the liquid oxygen bath at the base of the low pressure column, and caused a number of violent explosions.

The problem once diagnosed was remedied by placing silica or alumina gel filters in the rich liquid pipelines leading from the high pressure to the low pressure columns.   In some cases the reboilers were redesigned to ensure that they operated as through-flow systems, so that any hydrocarbons would leave with the product and not accumulate.   Also alternative air inlet ducts were provided so that the purest air was admitted to the compressors.   Stringent precautions against hydrocarbon accumulation are now standard in all air separation plant.

The insulation problem was recognised as a result of a number of mishaps. Non-inflammable insulants like mineral wool were considered safe until it was discovered that significant amounts of oil were being added to it by the manufacturers to facilitate handling.   At one time it was also usual to support major components within the cold box on balks of timber or to use a timber flooring in the cold box, but these also caused trouble.   The oxygen enriched liquid condensing adjacent to the top of the upper column soaked the mineral wool insulation, and progressively drained down by preferential paths to the base of the cold box thus penetrating the woodwork.   In addition to scrupulously avoiding the use of all inflammable materials in the cold box, it became good practice to purge the insulation with nitrogen.

The Linde Double Column has dominated air separation. BOC attempted to introduce a novel type of single column cycle (the Rescol cycle) for certain tonnage duties but it appeared not to be as flexible as the Double Column for the complex spectrum of products which increasingly was being demanded. In many parts of the world air separation plants were being required to produce oxygen both as gas and liquid (with a purity of 99.5%). Some of the nitrogen was also required as high purity product of which part was likely to be needed as liquid. There was also a growing market for argon, as well as for krypton and neon. So additional columns were attached to the Double Column and additional refrigeration was provided by recirculating nitrogen in an ancillary cycle with its own expander. The control of such installations to meet various product mixes became very demanding, and was met by computer based systems.

The story is still unfolding. A recent development has been 'front end clean-up'. Molecular sieves are now being used in special pressure-swing adsoption systems to remove water vapour, CO_2 and hydrocarbons from the air feed before it is cooled. In the L'Air Liquide system this is done by a combination of alumina gel for water vapour and molecular sieve for CO_2, the first being arranged in an annulus around the second, the flow direction being radial. The heating requirement for this type of adsorption is very modest since in combination with pressure swing it is only necessary to drive a temperature pulse through the beds rather than raising the whole bed to regeneration temperature. The resulting avoidance of switching of the heat exchangers substantially improves their performance and life, as well as allowing the column to operate more efficiently.

The design of the expansion turbines has been greatly refined, and Union Carbide engineers are now claiming isentropic efficiencies in excess of 90%. The overall thermodynamic efficiencies of these large plants are now better than 25%, and since some of the product is needed as liquid it is highly unlikely that the cryogenic route will be superseded for large scale production.

Low Temperature Separation of Other Gas Mixtures

(a) Coke Oven Gas, and other Coal-Derived Mixtures

By 1905 both Linde and Claude were involved in developing plants for the separation of hydrogen for coke oven gas in relation to ammonia synthesis. Due to the very low boiling point of hydrogen, which is far below the freezing points of carbon monoxide and nitrogen, the separation depended on partial condensation of the feed gas under refluxing conditions down to the lowest possible temperature short of freezing the CO. In the case of Linde plants this was achieved by the evaporation of a liquid CO/N_2 mixture and in the Claude plant by expansion of the product hydrogen stream in an expansion engine. The presence of a small amount of nitrogen in the product did not matter and the residual CO was removed by nitrogen scrubbing (Fig.8).

It is a tribute to the coppersmiths of that period that these plants were commissioned and operated over many years without an undue number of accidents. Large quantities of carbon monoxide, both as liquid and gas, were held up in the plant and constituted an enormous toxicity hazard. In addition hydrogen was present under pressure over a wide range of temperatures throughout the plant, and its low ignition temperature posed a grave explosion threat. Traces of peroxides were also present in some of the feed gases and these tended to condense in bands at an appropriate temperature in heat exchangers, with dramatic results.

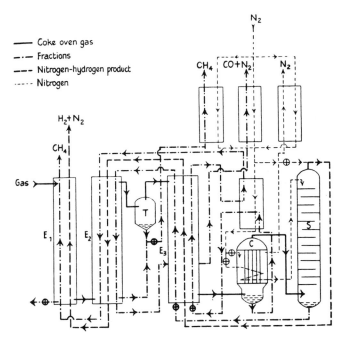

Figure 8.

A coke oven gas separation plant used to produce ammonia synthesis gas. Carbon monoxide is scrubbed from the product by liquid nitrogen in column S. Most of the methane is condensed out in C.

Incidentally a Linde designed coke oven gas separation plant was operating in Flixborough during the last war, and a methane-rich liquid sidestream from that plant enabled the first tests to be made on fuelling a bus with liquid methane. Few of these plants are now in use since hydrogen is produced by other routes.

(b) Ethylene from Cracker Gases

The earliest ethylene separation plants were built in the USA and USSR in the middle 1930's. Enormous growth has occurred since 1950 with plants producing high purity streams of both ethylene and propylene for polymer manufacture, together with less pure streams of both for chemical synthesis.

Early plants adopted petroleum practice as far as possible[3]. Thus a high pressure feed (at about 40 bar) was employed so that the separation temperatures could be kept above -90°C. The columns (Fig.9) were constructed from appropriate grades of steel and were spaced out in a row with separated condensers and reboilers, each individually lagged. A complex cascade refrigeration system provided the necessary cooling for the condensers and heat for reboilers. The resultant plant had good operating flexibility but was complex and of high capital cost.

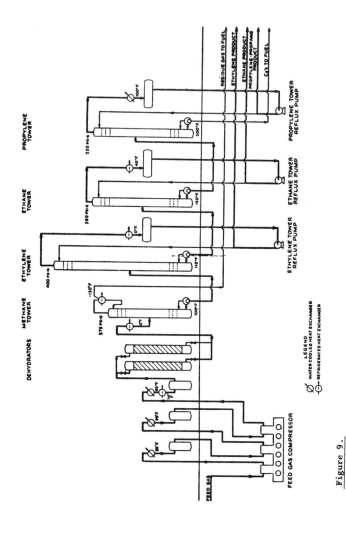

<u>Figure 9.</u>

An early cracker gas separation plant built using petroleum engineering practice.

One of the first ethylene plants built in Europe (Fig.10) was that designed and erected by Petrocarbon Ltd. at Partington near Manchester in 1947[4]. A feed pressure of only 10 bar was employed and the minimum temperature was -135°C. Separation was easier at the lower pressure, requiring lower reflux ratios, and aspects of air separation practice were adopted to save power and first cost. Thus there was a greater degree of integration of heating and cooling duties, copper and bronze were used as constructional materials at the lowest temperatures, and the low temperature columns were grouped in an insulated cold-box.

Within 10 years the scale of operation had grown and all the units became larger and more closely adapted to their particular function. The advantage of grouping the columns gave way to the benefits of accessibility.

Hence cracker gas separation plants became very different from other cryogenic separation plants. The high value of the feed compared to air made the balance of capital and operating costs quite different. The supporting cascade system enabled reflux conditions to be specified at will, and did not require ingenious integration with side-stream and product enthalpy changes. There was no 'pinch' in terms of refrigeration.

However the problems of optimising control loomed very large and could only be tackled by on-line computers. Most recently the methods of Second Law Analysis promoted by Linhoff[5] and fellow workers have been applied to these plants with resulting changes that are not unrelated to the classical low temperature approach.

Conclusion

In cryogenic gas separation the tremendous strides of the pioneers at the turn of the century have given way to steady growth and refinement. Improved compressors, heat exchangers, expanders and columns have cheapened first cost and reduced power consumption. These refinements will continue but the scope for radical improvement is small.

New circumstances will create new opportunities. An example is the Australian venture in which Cryoplants Ltd. has been involved in coupling the refrigeration demands required in the production of liquid oxygen and nitrogen with available 'cold' from the evaporation of LNG.

The success of low temperature gas separation is still based on the properties of the liquid/vapour interface. The fluxes achieveable in distillation are unlikely to be surpassed in any other separation process because all others involve the presence of either another phase or a diluent, both of which add diffusional resistance. The power required for low temperature gas separation could be further reduced but only at the expense of additional cost, and this is not yet justified.

The upgrading of lean natural gas and the bulk production of nitrogen for enhanced oil recovering are creating new challenges. The future of gas separation will continue to be interesting but it is unlikely to be scintillating.

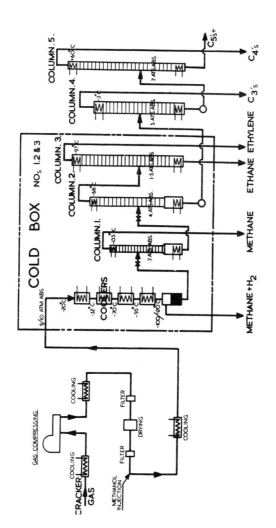

Figure 10.

An early ethylene plant designed by Petrocarbon Developments Ltd. using a cold-box approach and a low feed pressure.

References

1. Ruhemann, M.,   'The Separation of Gases', Clarendon Press, Oxford,
    Second Edition, 1949, p.154–162.

2. Kapitza, P., Jour.of Phys., 1939, 1, p.7–28.

3. Pratt, A.W. and Foskett, N.L.,   Low temperature processing of light
    hydrocarbons, Trans.Amer.Inst.Chem.Eng., 1946, 42, 149.

4. Ruhemann, M.,   The control of low temperature gas separation plants,
    Trans.Inst.Chem.Eng., 1952, 30, 125–141.

5. Linhoff, B., New concepts in thermodynamics for better chemical process
    design, Chem.Eng.Res.Des., 1983, 61, 207–223.

Pressure Swing Adsorption Methods

By N. F. Kirkby

DEPARTMENT OF CHEMICAL AND PROCESS ENGINEERING, UNIVERSITY OF SURREY,
GUILDFORD, SURREY GU2 5XH, UK

Introduction

Industry uses a wide variety of methods for separating mixtures of gases. A feature common to many of these methods is an additional phase, where the second phase, although often in equilibrium with the gas or vapour, has a substantially different composition. Most commonly, the second phase is a liquid; distillation and absorption are processes based on this principle. However, there are examples where the formation of an appropriate liquid phase can be a problem. In distillation it may be expensive to create and maintain the liquid phase, especially when the gas mixture has a very low dew point, or when the system possesses an inconvenient azeotrope, and in order to use absorption a suitable solvent must be found. Adsorption processes use a solid adsorbent in a similar way to the use of a solvent in absorption. Here, the second or 'adsorbed' phase is reversibly bound to the surface of the adsorbent.

In vapour-liquid methods, since both phases are fluid, it is relatively simple and very common to move the phases in opposite directions within the separation apparatus and thereby achieve a steady state operation. In adsorption methods, the solid phase is not usually moved because of the problems of transferring solid through gas tight seals and attrition, but moving bed systems are used and were proposed before 1920. Once a charge of solid adsorbent is to remain within the apparatus, non-steady state operation becomes inevitable. Two general schemes of operation are used: parametric pumping and parameter swing methods, and the two favoured parameters are temperature and pressure.

Parametric pumping and parameter swing methods are not always readily distinguishable but a crude distinction can be made as follows. In parametric pumping, the parameter fluctuations form moving waves which carry a concentration wave with them down the fixed bed. In parameter swing methods the parameter fluctuations are imposed with time but do not vary with axial position and the concentration waves are generated by simultaneous movement of the fluid phase. It is interesting to note that whilst parameter swing and parametric pumping methods are quite common, there are very few examples of the analogous processes based on the use of vapour-liquid equilibria.

For gas separation there are two common manifestations of parameter swing methods: thermal swing adsorption (TSA) and pressure swing adsorption (PSA). The former is the older process and still has wide application in natural gas sweetening and gas drying operations as well as in cryogenic air separation plant, where TSA is used to remove water, carbon dioxide and heat from the feed air.

The largest applications, by size of industrial plant, of pressure swing adsorption are hydrogen separations, but the largest number of PSA plants are used for the small scale production of oxygen from air. PSA is also used for air drying and has been used on a number of more exotic mixtures such as hydrogen isotope separation. PSA drying processes were sometimes called 'heatless drying', a name which emphasised the essentially adiabatic (and approximately isothermal) nature of all PSA processes. PSA processes therefore span a very wide range of scale and applications and are covered by thousands of patents.[1,2] The advantages of PSA are its simple and robust operation with rapid start up and shut down, and high product purity rivalling distillation in many applications. The disadvantage of PSA is that often the product yield is poor and therefore operating costs are high, especially at large scale. Many of the PSA patents describe process configurations which are claimed to improve yield or product purity or reduce power consumption or capital costs, although few of these are used commercially. This paper attempts to trace the development of these processes and to present a systematic approach to categorising them but will ignore the theoretical modelling. Firstly, however, some mention of adsorption must be made because the development of PSA processes has been stimulated and sustained by the development of suitable adsorbents.

'Adsorption' as Used in Pressure Swing Adsorption

The commonly used adsorbents are all highly porous solids and are mainly
either zeolites[3] or forms of carbon although silica gel and alumina are
still used for some applications. The role of the adsorbent is to retain an
adsorbed phase with a different composition from the bulk gas phase. This
may be achieved in one of two ways: the selective adsorbent retains an
adsorbed phase which is at or near thermodynamic equilibrium with the bulk
phase. The molecular sieve has a pore size distribution so that certain
molecular species can gain access to the adsorption sites more quickly than
others. The amount and composition of the equilibrium adsorbed phase is a
function of temperature and the partial pressures of the components. The
rate of mass transfer between the gas and adsorbed phases is normally
controlled by a resistance, which may be external mass transfer, pore
diffusion or occasionally a rate of adsorption. Pore diffusion may be
subdivided further since most adsorbents have bimodal pore size
distributions. The larger pores (macro pores) are generally the cracks and
boundaries between the crystallites which make up the adsorbent pellets. In
molecular sieves, the small pores within the crystallites or between
graphitic planes usually control the mass transfer rates of individual
components.[4]

The important fact to note is that it is not sensible to describe any of
these solids as 'selective adsorbents' or 'molecular sieves' in isolation,
both terms are descriptions of the gas–solid interaction. Some examples,
particularly pertinent to PSA, should illustrate the point.

5A zeolites are commonly used in PSA to separate oxygen from air at room
temperature and modest pressures. Under these conditions the equilibrium
between the gas and the adsorbed phase is achieved very rapidly and the
zeolite is operating as a selective adsorbent. Typical isotherms[5] are shown
in Figure 1. Nitrogen is the more adsorbable component and therefore the
gas phase becomes enriched in oxygen. However, if the process is operated
at much lower temperatures, although nitrogen continues to be the more
adsorbable component at equilibrium, the diffusivities of both oxygen and
nitrogen are reduced so that the dominant operation is molecular sieving.
Oxygen can gain access to the adsorption sites more quickly than nitrogen;
therefore the gas phase becomes enriched in nitrogen.[6]

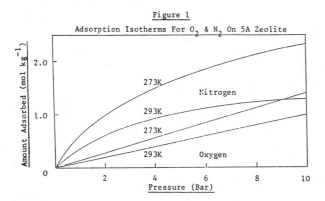

Figure 1

Adsorption Isotherms For O_2 & N_2 On 5A Zeolite

There are no selective adsorbents which adsorb more oxygen than nitrogen and are commercially viable for the production of nitrogen from air. Two options remain; the usual method is to use a carbon molecular sieve (CMS) which will operate economically at room temperature. In the Bergbau -Forschung CMS the equilibrium isotherms for oxygen and nitrogen are very similar but oxygen is adsorbed much more quickly [7,8] (see Figure 2). The alternative method is to modify the process configuration so as to produce a stream enriched in the more adsorbable component (see below).

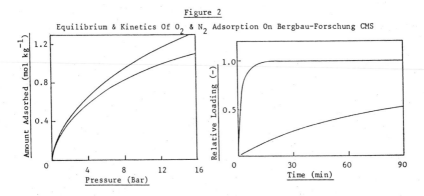

Figure 2

Equilibrium & Kinetics Of O_2 & N_2 Adsorption On Bergbau-Forschung CMS

The ideal adsorbent for PSA is one in which the selective adsorption and molecular sieving effects complement each other. Unfortunately, this situation does not often arise, despite nearly a century of research into the synthesis of adsorbents. In the last twenty years considerable research effort has been devoted to improving the process engineering of PSA, i.e. to devise process schemes which use the available materials to better effect.

The Fundamental Steps of PSA

As mentioned above, parameter swing adsorption processes involve the discontinuous operation of fixed beds of adsorbent. A sequence of batch operations is executed repeatedly, with several beds operating out of phase so that product gas is made at all times. Three steps are common to almost all pressure swing cycles: pressurisation, product release and de-pressurisation (see Figure 3).

Pressurisation normally involves the introduction of feed gas to the bottom of a bed whilst the top of the bed is closed. For the normal sizes of pellets (300μm to 3mm) and pressurisation rates ($\approx 10^3$–10^5 $Nm^{-2}s^{-1}$) the bed often has to be retained to prevent fluidisation. Pressurisation creates a region of gas adjacent to the closed end of the bed which is enriched in the less adsorbable component. Product release displaces this zone from the bed at constant pressure by introducing more feed at the bottom and simultaneously removing a similar amount of gas from the top of the bed. Depressurisation then proceeds with the top of the bed shut and gas is withdrawn or allowed to discharge from the bottom of the bed.

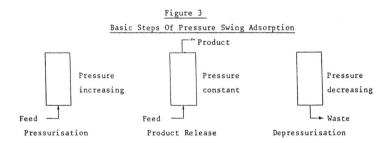

Figure 3

Basic Steps Of Pressure Swing Adsorption

Thus these three steps split a feed stream into two streams. For a given feed mixture the adsorbent is usually selected so that the desired product is the least adsorbable component, and this is the predominant component in

the product release stream, hence the name. The gas discharged during depressurisation contains the desorbing components and normally constitutes
the waste stream.

By definition, all pressure swing processes have at least a pressurisation
and a depressurisation step; indeed the first true PSA process consisted of
just these two steps and was patented by Finlayson and Sharp in 1932.[9]
Hitherto, desorption had always been in part effected by an increase in
temperature. However, no matter how good the adsorbent, these three steps
alone do not constitute a commercially viable process, and the Finlayson
and Sharp patent was not exploited.

There are two fundamental problems with the basic cycle: both the product
purity and the yield are poor. The purity may be improved by increasing the
pressure swing but this in turn may adversely affect the yield. To overcome
these problems a number of additional steps have been included in pressure
swing cycles. These extra steps can usually be classified according to one
of two roles which they perform in the cycle: to improve the regeneration
of the bed prior to the separation steps, or to increase the amount of
product quality gas released per cycle. Further regeneration steps improve
product quality by allowing the bed to start the separation in a cleaner
state. Enhanced product release steps allow more of the product rich zones
of gas to be released, and therefore enhance yield. Some of these steps
will be described in more detail below but it should be noted that these
roles are not always independent. For most PSA processes of fixed
configuration increasing the product amount per cycle reduces the product
purity. By adding steps to the basic cycle the process engineer seeks to
increase product purity whilst maintaining product amount or vice versa.
Alternatively, the engineer may try to maintain product amount and purity
whilst reducing the total pressure swing as this will increase the yield.

Further Regeneration Steps

The most common step in this category is the purge step. Some of the
product gas is expanded to the lower operating pressure and introduced at
the top of the bed, with the bottom of the bed open to the waste line as
shown in Figure 4. Whilst the operation is at constant total pressure, the
introduced gas contains the more adsorbable component at a very low partial
pressure and as a result further desorption occurs. This counter-current

arrangement (with respect to pressurisation) was used in a number of early TSA cycles as a means of heating the bed. Later Kahle used the step in an adsorption process which did not use either a pressure or temperature swing; instead the purge gas came from a separate gas supply.[10] Skarstrom, however, was the first to use a purge step in a PSA cycle, and the dramatic improvement in performance led to the first commercial applications of PSA.[11] The purge step is now a common feature of PSA processes and is used irrespective of whether the adsorbent is a selective adsorbent or a molecular sieve. The simple purge cycle is still used in many gas drying applications.

Figure 4

Further Regeneration Steps

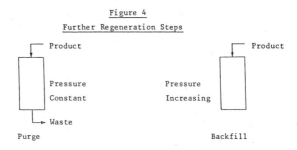

Purge   Backfill

The problem with the purge step is that it contributes gas to the waste stream. However, this is compensated for by the improvement in product concentration up to a particular optimum purge amount beyond which yield and product concentration both deteriorate. The purge step has now been replaced by or supplemented with a backfill step in many applications. Backfill is a partial repressurisation with product gas admitted at the top of the bed, with the bottom of the bed closed (see Figure 4). The effect of a backfill step is similar to that of a purge step, i.e. the top portion of the bed at least is cleaned prior to the pressurisation from feed. However, backfill has the advantage that no extra gas is contributed to the waste line and therefore product concentration is maintained with less penalty on yield than occurs with a purge step. Backfill also has a minor advantage in terms of energy consumption in that the subsequent pressurisation does not commence at the lower operating pressure.

Purge and backfill steps have slightly different and complementary roles in the improvement of bed regeneration and this accounts for the inclusion of both steps in a number of commercial systems.[12]

The basic form of the purge and backfill steps as described above have been modified in a few cycles. For instance, purging at the upper operating pressure with recompressed waste gas has been proposed for cycles where the more adsorbable component is the desired product, e.g. nitrogen from air. The backfill step has a number of interesting variants especially but not exclusively where molecular sieving is the primary mode of separation. In some cycles the bed is backfilled from the bottom instead of the top end, and in others gas is admitted at the top and bottom simultaneously.

Enhanced Product Release Steps

In specifying the mass balance to be required from a normal product release step the designer has to be able to predict how the concentration at the outlet from the bed will vary with time. A simple and neat solution to this difficult theoretical problem is to use a second-cut step[13] where a bed at high pressure donates gas to a second bed at high pressure, rather than directly to the product line (Figure 5).

Figure 5

Enhanced Product Release Steps

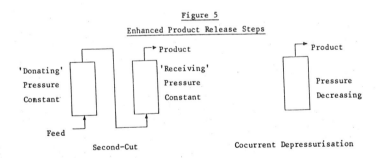

The second-cut donation step comes after normal product release and before depressurisation, and its timing does not have to be exact. It can commence some time before the critical breakthrough of the more adsorbable component and last until this breakthrough is complete, thus ensuring that the complete adsorptive capacity of the bed has been used. The accompanying step of receiving second-cut usually comes before normal product release and after pressurisation. The second-cut process improves yield but imposes a slight energy cost since some gas is effectively passed through a double length of bed. Second cut also increases the complexity of the pipework.

An alternative step is to partially depressurise the bed into the product stream after the normal product release. This cocurrent depressurisation step (Figure 5) allows a more complete release of the gas remaining at the top of the bed which is rich in the less adsorbable component. The gas released by this step is often used exclusively as purge or backfill gas. However, in some cycles where cocurrent depressurisation is used after a normal product release which is very short or where normal product release is not used at all, the initial release of gas can pass straight to the process product. Cocurrent depressurisation is a very attractive step in many applications because it can dramatically improve yield. The amount of product released is only part of this improvement; the other benefit is that the subsequent depressurisation to waste starts at a pressure below the upper operating pressure and thereby reduces the amount of waste gas.

Normally cocurrent depressurisation is combined with a backfill step on another bed. This combination of steps is sometimes called a bed pressure equalisation (BPE) step (even when the steps are stopped before the pressures in the donating and receiving beds are equal). On multi-bed cycles there can be a quite complex cascading of BPE steps, so that the eventual depressurisation to waste represents only a small fraction of the overall pressure swing. BPE steps are not always 'top to top' exchanges of gas; numerous variants have been patented, and in many cases it is not clear what effect the step has on the states of the donating and receiving beds.

There is much more which could be said about all the above steps and, for instance, how their roles vary according to the composition of the feed gas mixture and the type of adsorbent. Two general points will have to suffice: firstly, it should be obvious by now that the design of a multi-bed multi-step PSA cycle is not a trivial exercise, and secondly, although tempting, it is not entirely useful to compare the gas donated to steps such as purge and backfill with the reflux stream in conventional distillation.

The reflux analogy has been used on, for instance, simple purge cycles, to state that there is a minimum or critical purge to feed ratio, which depends only on the overall pressure swing, similar to the concept of a minimum reflux ratio. However, in many circumstances the optimum product concentration is achieved at a critical purge amount which depends on a much wider range of process variables than just the pressure ratio.

Also, in PSA, the recycled gas may be passed directly from one bed to another or via an intermediate storage vessel. In the latter case any concentration fluctuations with time in the gas from the donating bed may be averaged out before the gas enters the receiving bed. In direct transfer, breakthrough from one bed will create a distinct zone in the receiving bed, which in turn will affect the operation of subsequent steps in the cycle. This time dependent nature of the gas recycled in PSA systems is another reason for being cautious of the reflux analogy.

Trends in the development of PSA

Many of the early PSA processes used large pressure swings which were wasteful of energy and gave poor yields. As better adsorbents became available the pressure swings were reduced. In order to continue the reduction of power consumption and improvement of yield, more complex cycles were devised. This progress is limited by the balance between the capital costs incurred by the process complexity and the reduced operating costs.

Multi-bed cycles were introduced to improve the efficiency of separations where only one component of the feed mixture is the desired product, and are in effect separations of binary gas mixtures. However, there has been recent speculation that true multi-component separations can be performed with the different components in reasonable purities, being released separately from the beds at the appropriate stages in the operating cycle.[14] The design and operating complexity of such processes is not covered in this paper.

Another trend has been the use of more than one type of adsorbent within a single PSA plant. Some of the early processes used separate guard beds, and later guard layers within each bed. For example, in oxygen production from air some 5A zeolites were found to be sensitive to the moisture content of the feed air. Initially, an entirely separate PSA air dryer was added to the feed line but this effectively doubled the number of beds and valves. Later to reduce the number of valves, each zeolite bed received feed through its own separate bed of alumina, but ultimately a layer of alumina was added to the bottom of the zeolite beds. This final arrangement is in effect a PSA system designed to purify one component from a ternary mixture. Similar adaptions have been patented in order to produce argon from air,[15] and in time these multi-adsorbent processes will become more widely used.

A question which has exercised a number of workers in PSA over the last thirty years concerns the rate at which a given process can be cycled. It was immediately obvious that the adsorbent particle size would be a crucial factor when trying to cycle PSA processes very rapidly. Large particles would allow high flowrates for a given pressure drop but would limit the mass transfer rates so that, for instance, a selective adsorbent may start to act as a molecular sieve. Smaller particles give higher mass transfer rates but increase the pressure drop. This latter case has been elegantly exploited in a process described by Jones and Keller.[16] One way of looking at the Jones/Keller process is to say that the bed length, particle size, and cycle time have been selected so that several valves per bed in a conventional PSA process are no longer required. However, since the operation of the process now depends crucially on the motion of a pressure wave in the bed it could also be classified as a parametric pumping process. The inventors acknowledge the hybrid nature of their process by calling it 'pressure swing parametric pumping'. Rapid cycling improves adsorbent productivity, is consistent with the general trend towards process intensification, and could be of great future importance.

Finally, the theoretical modelling of PSA processes[17,18] has attracted relatively little attention and although considerable progress has been made, few models have been developed which accurately predict the performance of cycles in commercial use. Current models tend to be a compromise between accuracy and CPU time requirements. However, if the current trends in the cost, size and speed of computers continue, it should not be too long before accurate models become widely available. Theoretical models offer the long term hope that adsorbents and cycle configurations may be selected on a more rational basis and that cycles may be optimised to meet fully the demands of an increasingly competitive market.

References

[1] D.Tondeur and P.C.Wankat, Sep. Pur. Methods, 1985, 14(2), 157.

[2] H.Lee and D.E.Stahl, A.I.Ch.E. Symp. Ser., 1973, 69(134), 1.

[3] D.W.Breck,'Zeolite Molecular Sieves', Wiley-Interscience, New York, 1974.

[4] D.M.Ruthven,'Principles of Adsorption & Adsorption Processes', Wiley-Interscience, New York, 1984.

[5] N.F.Kirkby, Ph.D. Thesis, University of Cambridge, 1983.

[6] A.Arkharov, G.Voronin, M.Dubinin, I.Kalinnikova, Y.Nikiforov, V.Serpinsky and N.Fedoseeva, XV Int. Cong. Refrig., 1973, A3-3.

[7] P.Leitgeb, Linde Reports on Science and Technology, Oct. 1975.

[8] H.Juntgen, K.Knoblauch and H.J.Schroter, Ber. Bunsen-Gesell. Physik. Chemie, 1975, 79, 9, 8240826.

[9] D.Finlayson and A.J.Sharp, Br. Patent 365092, 1932.

[10] H.Kahle, Chem. Ing. Techn., 1954, 26, 75.

[11] C.W.Skarstrom, U.S. Patent 2944627, 1960.

[12] N.F.Kirkby and C.N.Kenney, Fundamentals of Adsorption, Engineering Foundation Conference, Santa Barbara, May 1986, in press.

[13] J.W.Armond, Spec. Publ. Chem. Soc., 1980, 33, 92.

[14] S.Natarij and P.C.Wankat, A.I.Ch.E. Symp. Ser., 1982, 219(73), 29.

[15] J.W.Armond, Australian Patent 515010, 1981.

[16] R.L.Jones and G.E.Keller, J. Separ. Proc. Technol., 1981, 2(3), 17.

[17] N.F.Kirkby and C.N.Kenney, Chem. Eng. Sci., in preparation.

[18] J-L.Liow, Ph.D. Thesis, University of Cambridge, 1986.

Priestley Lecture
On the Science of Deep-sea Diving - Observations on the Respiration of Different Kinds of Air

By E. B. Smith

OXFORD HYPERBARIC GROUP, PHYSICAL CHEMISTRY LABORATORY, SOUTH PARKS
ROAD, OXFORD OXI 3QZ, UK

In 1776 Joseph Priestley performed his critical experiments on the respiration of air and clearly identified it as a 'phlogistic process'.[1] These investigations were the first steps in the understanding of the oxidative processes involved in metabolism. Some four years earlier Priestley had read before the Royal Society his substantial and important paper : 'Observations on Different Kinds of Air'.[2] In this he describes the identification of nitric oxide and hydrogen chloride as well as the preparation of 'fixed air' (carbon dioxide). The investigation of 'airs' was to be the main theme of Priestley's work in the years that followed and led to the discovery of 'dephlogisticated air', oxygen, in 1774.

The discoveries I want to discuss today result, curiously, from a bringing together of these two themes, respiration and 'different kinds of air'. Indeed a most appropriate title for this lecture would be: 'Observations on the respiration of different kinds of air'. The research concerns respiration not of normal atmospheres but rather of exotic gases. Long after the biological role of oxygen had been elucidated it was found that even the most chemically unreactive gases could have unusual and dramatic effects when respired. These discoveries and the manner in which they were made would have been readily appreciated by

Joseph Priestley.   First they required an interdisciplinary approach that was more characteristic of the research of Priestley's time. Second they were discovered in the absence of a framework of scientific ideas and indeed their discovery would have endorsed his 'Baconian conviction' that only facts are important.   It is a field where serendipity has played a more important role than theoretical understanding ; it is a field that still awaits its Lavoisier!   A third factor which would have pleased Priestley is that it is a subject of very direct practical importance. He took great pleasure from his discovery of a practical way to make 'Pyrmont' water by artifically carbonating water.   He wrote 'what cost you five shillings will not cost me a penny'. The practical application of the 'respiration of different kinds of air' include general anaesthesia and deep sea diving.   Surprisingly these two subjects are not unrelated.

Diving

Almost three-quarters of the surface of the earth is covered by water and it is natural that from the earliest times man would wish to gain access to these regions.   It is claimed, on the basis of archaeological studies, that Neanderthal man dived for food. By 4500 BC diving was well established and from the times of the ancient Greeks diving for sponges has been an recognised profession.   The Greeks are said to have established laws governing the rights of divers to salvaged goods; the share increased with the depths from which the goods were recovered. Even today breath-hold diving plays an important role in Asian pearl diving communities.   It is a highly developed skill with many records of divers remaining under water for over 4 minutes. However without specialised equipment man's capacity to explore beneath the ocean is severely limited.

The first use of diving equipment is usually attributed in legend to Alexander the Great who was said to have been lowered into the sea at the Straits of the Bosphorus in a glass barrel during the 3rd century BC (Fig.1). The medieval portrayal of his adventure shows the sealed glass barrel with lamps burning within, suggesting that Alexander was lucky to survive! More practical development of the diving bell occurred in the 16th century; in the mid 17th century bells were used to recover the cannon (which each weighed 1 ton) from the Swedish warship Wasa sunk in 30 m of water. In 1691 Edmund Halley improved bell technology by devising a method of replenishing the air supply using weighted barrels. In the 19th century the traditional diving suit was developed - a closed suit with air supplied under pressure by a pump on the surface. This was to provide the basic technology

1. A medieval view of Alexander the Great being lowered into the Straits of the Bosphorus in a glass barrel.

of diving until the development of SCUBA gear by Cousteau and Gagnan in 1943. In recent years a sophisticated technology for saturation diving, in which divers remain at pressure for periods of many days or even weeks, has been developed. The scientific principles underlying modern diving technology will be the subject of this lecture.

Before moving to a consideration of the more subtle factors that place restraint on human undersea activity, it might be helpful to identify two very basic constraints. First, for the free diver the pressure inside and outside the diver must be essentially equal, otherwise his lungs would collapse. Divers must be supplied with air equal to their environmental pressure which increases by 1 atm. for every 10 m they descend below the surface. Second, the divers cannot breathe pure oxygen except in very shallow dives since oxygen has a number of adverse physiological effects at pressures greater than say 2 atm and has the capacity to induce convulsions. These considerations mean that divers must be supplied with a gas mixture containing both oxygen and a dilutant gas or gases at pressure. This sets the context for deep diving research.

There is a further factor that can limit deep diving – decompression sickness. As divers breathe gases under pressure more gas dissolves in the body, according to Henry's Law. If a diver returns to the surface too quickly this excess gas can be released and can cause pain and even paralysis or death. The Hyperbaric Group at Oxford has been very active in investigating the problems of decompression sickness using ultrasonic detection equipment developed by my colleague Dr. S. Daniels. However in this lecture I will confine my attention to those problems that men face while at depth rather than those which restrict their return to the surface.

From early times attempts have been made to develop an alternative strategy for undersea exploration - pressure vessels and pressure suits. Recent advances have enabled divers in pressure suits to reach considerable depths but it is generally accepted that the development of underwater resources requires the involvement of free divers, particulary in emergencies.

First Surprise : Inert Gas Narcosis

As divers equipped with the reliable standard diving suit pushed to depths greater than about 150 ft and better procedures to avoid decompression sickness were developed, a new problem occurred.   The divers became confused and euphoric and eventually, as they proceeded to greater depths, could lose consciousness. In 1935 it was suggested that this could be due to a narcotic effect of the nitrogen in the air but this view was not generally accepted for many years. Cousteau romantically referred to this phenomenon as 'rapture of the deep' but the phenomenon is now generally known as 'inert gas narcosis'. It sets severe limits on air diving - 185 ft for British naval divers. It is now understood that inert gas narcosis is a manifestation of incipient general anaesthesia - a state that can be induced by all non-reactive gases (which have no more specific adverse effects). Thus even a gas as inert as xenon can cause narcosis and has in fact been used clinically as a general anaesthetic.

The most striking feature of anaesthetic substances is that they have no structural features in common (unlike many series of compounds that are centrally active in the nervous system). Although anaesthetics can affect many neural areas - central and peripheral - the mechanism by which they cause loss of consciousness appears specific. The effective anaesthetic doses may vary over a wide range, as given for mice in Table 1. The anaesthetic pressures for other mammals and humans are very comparable.[3] The lack of structural features leads one to suppose that the relative potency of anaesthetics is governed by their intermolecular forces and it is natural to assume that anaesthetic potency is controlled only by very general physico-chemical principles. Underlying almost all current investigations is the so-called unitary hypothesis, which proposes that all general anaesthetics act by the same mechanism.

Table 1   Partial pressures (in atm) required to produce general anaesthesia in mice

N_2	33	C_2H_2	0.85
CF_4	19	CF_2Cl_2	0.40
C_2F_6	18	CH_3CClF_2	0.25
Ar	15	CHClF	0.16
SF_6	6.1	$c-C_3H_6$	0.16
Kr	4.5	$(C_2H_5)_2O$	0.030
N_2O	1.5	$CHCl_3$	0.0084
Xe	0.95	Halothane	0.0077
C_2H_4	1.4	Methoxyflurane	0.0022

Ferguson's principle

Not surprisingly, in view of the unitary hypothesis, much of the speculation about the mechanism of general anaesthesia has come from physical chemists. The most general of these approaches is that of Ferguson.[4] Ferguson's principle states that although the equilibrium concentrations of various anaesthetics required to produce a chosen level of anaesthesia may vary widely, the thermodynamic activities corresponding to those concentrations lie in a relatively narrow range. This iso-narcotic activity (defined relative to the pure liquid as standard state) for a number of common gaseous anaesthetics is found to be approximately 2×10^{-2}. At equilibrium the thermodynamic activity of the anaesthetic will be the same in all regions within the central nervous system including the (as yet unidentified) site of action. Though the concentrations at different sites will differ due to different molecular characteristics, the activity at all sites would be the same. One serious implication of this argument is that if all anaesthetics were equally potent under conditions of equal activity, it would be impossible to obtain any information about the nature of the site of action of anaesthetics by studying the variation of potency with molecular properties. It is, therefore, important to examine carefully the validity of Ferguson's principle.

The general limitation of the principle becomes most apparent with gaseous anaesthetics; it is equivalent to relating anaesthetic potency to ideal solubility which any solute is independent of the nature of the solvent. It is therefore necessary to consider only the intermolecular forces between the anaesthetic molecules, i.e. the solute molecules, and one can ignore those between the anaesthetic and its site of action. Ferguson's principle may be written as:

$$a_{narc} = P_{narc}/p^0$$

where a_{narc} is the activity of an agent required to cause anaesthesia, P_{narc} the equivalent partial pressure of the agent, and p^0 its vapour pressure at the temperature of the experiment. Thus;

$$\log(p_{narc}) = -\log(1/p^0) + \log(a_{narc})$$

A plot of $\log(p_{narc})$ vs $\log(1/p^0)$ should yield a straight line of unit negative slope. $1/p^0$ is the ideal solubility of the anaesthetic at one atmosphere partial pressure. If concentration at a particular site, rather than the activity, is the relevant variable then we would expect Ferguson's principle to hold only for agents whose solutions at the site of action are ideal or deviate from ideality in a constant manner. To test Ferguson's hypothesis we must investigate the anaesthetic potency of substances of unusual solubility properties.

Fluorine compounds

The unusual behaviour of fluorine compounds, which has been the subject of considerable study by those interested in solubility, was suggested as a means of characterising the site of action of general anaesthetics. The phase in which anaesthetics act can be identified only if special attention is directed, not to the general properties of anaesthetics as in Ferguson's approach, but to situations in which the specific nature of the interaction between the molecules of the anaesthetic and the molecules of the 'solvent' area is manifest.

Such situations are provided by systems which deviate markedly from the ideal. Generally, fluorinated compounds have relatively weak intermolecular forces and these lead to large positive deviations from ideality in mixtures with more typical nonpolar solvents. The largest deviations are observed when the solvent has strong intermolecular forces. Thus CF_4 and SF_6 have the lowest known solubility in water, whereas their behaviour in nonpolar solvents, though anomalous, is less exceptional. CF_4 and SF_6 are anaesthetic at a thermodynamic activity that is greater than the activity of most common anaesthetics by a factor of 10 - a striking departure from Ferguson's rule. This departure from the Ferguson hypothesis suggested that fluorine compounds might provide a key to characterising the site of action of general anaesthetics.

Hydrate theories

In 1961, Pauling[5] and Miller[6] independently suggested that anaesthetics act in the aqueous phases of the central nervous system. Though the two theories differ in detail, both seek to relate anaesthetic potency to the stability of the gas hydrates which many anaesthetics can form in aqueous solution. To account for the fact that such hydrates are not stable under physiological conditions, Pauling suggested that they might be stabilised by the charged side-chains of proteins in the encephalonic fluid. Hydrates, once formed, could increase the impedance of the neural network or occlude pores in membranes. They could also impair the reactivity of enzymes by a 'cage' effect. Miller did not invoke the actual presence of hydrates within the body but considered the effects of ordering which simple solutes are supposed to induce in neighbouring water molecules (the 'iceberg effect'). He suggested that these 'icebergs' surrounding anaesthetic molecules could impair neural function in much the same manner as the hydrates proposed by Pauling.

Both proponents suggested that a suitable test of their theories would be to determine the degree to which the anaesthetic potency of a substance is related to the stability of its hydrate as represented by the reciprocal of its dissociation pressure at 0^0C. The examination of this relationship for a wider range of anaesthetics shows that it is far from satisfactory (Fig.2).[7]

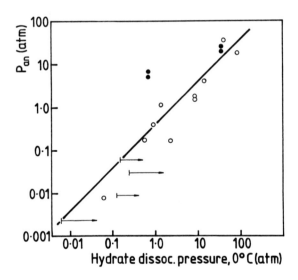

2. The correlation of anaesthetic pressures with hydrate dissociation pressures at 273K. The arrows indicate anaesthetic substances that do not form hydrates at their saturated vapour pressures. The filled circles represent fluorine compounds.

SF_6 would be five times more potent as an anaesthetic if its hydrate stability were an accurate guide. The credibility of the theory is further reduced by the fact that it has proved impossible, to date, to make hydrates of a number of anaesthetic substances, including both ether and halothane.

Advocates of the hydrate theory suggested a further test. If mixtures of gases which form the so-called class 1 hydrates and those which form class 11 hydrates were to be used to induce anaesthesia, then a positive synergistic effect would be expected as a mixed hydrate could very readily be formed. This hypothesis was tested by Eger et al.[8] in a study of the anaesthetic potency of halothane-xenon and halothane-ethylene mixtures. The potency was found to be that expected if the effects of the two anaesthetics were simply additive. These studies show that there is essentially no physico-chemical evidence to suggest that the aqueous phases of the central nervous system are the site of anaesthetic action.

Lipid solubility theory

At the end of the 19th century, Meyer[9] and Overton[10] noted the striking correlation between anaesthetic potency and fat solubility for a wide range of anaesthetic substances. On the basis of this observation, they advanced the lipid solubility theory. The theory has been formulated in modern form by Meyer[11], 'Narcosis commences when any chemically indifferent substance has attained a certain molar concentration in the lipid of the cell. This concentration depends on the nature of the animal or cell but is independent of the narcotic.'

This model predicts that anaesthetic potency should be directly proportional to solubility at the site of action. This relationship has been carefully tested using olive oil and other solvents selected to match the properties of the site of action. For olive oil the correlation between the solubility of anaesthetic substances and their potency was shown to hold with remarkable accuracy and is predictive to ± 20 per cent over a potency range of about a factor of 10^5 — one of the more remarkable correlations in science (Fig.3). Unfortunately, olive oil is not too well characterised from a physico-chemical standpoint and it is desirable to use simpler substances in order to establish the physical nature of the area in which anaesthetics

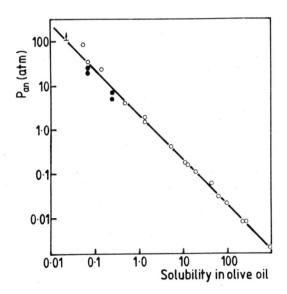

3. The correlation of anaesthetic pressures with solubility in olive oil. The anaesthetic potency of a substance is seen to be directly proportional to its solubility in this solvent. The filled circles represent fluorine compounds.

act. The most suitable property with which to characterise
solvents is the solubility parameter δ, defined by:

$$\delta = \left[\frac{-E^{vap}}{V} \right]^{\frac{1}{2}}$$

(where E^{vap} is the energy of vaporisation of a substance and
V is molar volume) which is rough measure of the strength of the
intermolecular forces within the solvent. The correlation of
anaesthetic potency with solubility in a series of solvents with δ
in the range 6-10 $(cal\ cm^{-3})^{1/2}$ has been examined.[12] For most
anaesthetics a fair degree of fit can be obtained in all cases,
but the fluorinated molecules deviate in one manner in the
solvents of low δ and in the opposite direction with solvents of
high δ. Minimisation of these deviations indicates that the
solubility parameter of the area in which anaesthetics act is
9 $(cal\ cm^{-3})^{1/2}$. Essentially this is a region with
intermolecular forces of the general strength of those in benzene.
A corollary is that we can estimate the concentration of
anaesthetic at the critical site of action that is required to
produce anaesthesia as 0.05 mole $litre^{-1}$. Mullins[13] has advanced
a modified version of the lipid solubility theory which proposes
that narcosis commences when a certain critical volume fraction of
an inert substance is attained at the hydrophobic site of action.

The Molecular Nature of the Site of Action of Anaesthetics

    It is now generally accepted that the site of action of
anaesthetics is hydrophobic, but the molecular nature of the site
is still a matter of debate. The traditional view is that
anaesthetics act in the lipid region of the cellular membrane, to
which is often added the belief that the induced perturbation
interferes with ion transport through the membrane. However,

studies of the effects of anaesthetics on membrane processes lend little support to this view. Membranes are surprisingly resistant to the effects of anaesthetics, and the changes that result from the application of anaesthetics could easily be produced by variations of temperature within the physiological range. In 1954, Johnson, Eyring, and Polissar wrote[14], "... it is not necessary to assume that anaesthetic action involved cell lipids." Indeed, on the basis of the results of experiments on the action of anaesthetics on ion transport processes, we concluded some years ago that "the results give no indication that the general perturbation of membrane processes is an essential feature of the mode of action of general anaesthetics at clinical concentrations."[15] There seems no reason to modify this conclusion at the present time.

Possible alternative sites of action of anaesthetics are hydrophobic regions of proteins. In recent years considerable efforts have been devoted to evaluating this possibility. For most of the proteins investigated, no functional change was observed on the application of anaesthetics. However the light output of luminous bacteria is found to be inhibited by clinical concentrations of anaesthetics and this inhibition appears to arise from a specific interaction of the anaesthetics with luciferase, the enzyme responsible for light emission[16]. This suggests that for this system a hydrophobic site within an enzyme may be the critical site of action of the anaesthetic molecules. This does not conflict with the Meyer-Overton model since the interaction appears to be closely related to solubility in hydrophobic solvents. This has been confirmed by the recent experiments on firefly luciferase by Franks and Lieb.[17] They showed that, over a 100,000-fold range in potency, the anaesthetic concentrations required to reduce the activity of purified firefly luciferase by 50% were essentially identical to those which induce anaesthesia in mammals. Anaesthetics were shown to act competitively with the luciferase substrate. Franks and Lieb concluded that the volatile general anaesthetics, despite their diversity, all acted by competing with the endogenous ligands by binding to specific receptors, and it was suggested that the anaesthetic binding site can accommodate two small anaesthetic molecules but only one molecule of anaesthetics larger than hexanol.

Though the detailed mechanism of action of general anaesthetics has yet to be established, the strong correlation of anaesthetic potency with solubility in fatty solvents suggested that divers could avoid inert gas narcosis by breathing less soluble gases such as helium. The use of helium in diving was in fact suggested by Hildebrand and others in the 1920's as a possible help in overcoming decompression sickness (inert gas narcosis was not identified at that time). Following many years of development, diving with helium was brought into service in spectacular circumstances in 1939. The U.S. submarine 'Squalus' sank beyond the reach of air divers. The new experimental helium technique was brought into play and all the crew were saved and the vessel salvaged.

The Second Surprise : The High Pressure Neurological Syndrome

As divers using helium went to even greater depths in the 1960's new problems arose. At about 200 m divers were observed to experience new symptoms quite different from those of inert gas narcosis: tremors, nausea, numbness etc.[18] Similar effects were also observed in mice and other experimental animals exposed to high pressures of helium.[19] Somewhat higher pressures than those to which divers have been exposed can induce convulsions in animals.

At first it was not clear if these effects were due to special properties of the helium used in the diving mixture or due to pressure per se. The classic work of Regnard[20] and others showed that with aquatic animals the first effect of pressure is a stimulation of the central nervous system at pressures above 50 atm. At higher pressures (200-300 atm), spontaneous muscle contraction occurs and the animals become paralysed. Still higher pressures (400 atm) prove lethal. The interpretation of experiments performed on aquatic or amphibious animals is unequivocal since the hydrostatic pressures may be applied in the absence of potentially narcotic gases; however, in experiments with mammals it is not always easy to distinguish between the effects of pressure per se and the narcotic effects of the gases breathed. In experiments with the amphibious Italian great crested newt the same response was observed if pressure was applied using helium (or indeed neon) or applied hydrostatically. Thus it was concluded that helium, and probably neon, acted essentially as pressure transmitting agents and that if these gases were anaesthetic, their anaesthetic pressures were greater than the tolerable mechanical pressure.[21] This indicated that the symptoms observed in men and animals breathing helium-oxygen mixtures at high pressures were due directly to the effects of pressure; the effects are now known as the high pressure neurological syndrome (HPNS) which constitutes a serious barrier to diving to great depths.

The Third Surprise

It was found that the symptoms of HPNS were less severe if slow compressions were employed and using this technique divers reached depths of well over 1,000 ft. Nevertheless it appeared that the effects of pressure per se would set the ultimate limit on the depths at which divers could work. However there was a third surprise awaiting research workers in this field - the fact that pressure and anaesthetics appear to be mutually antagonistic. The first observation of this remarkable effect was made by Johnson and Flagler in 1951[22]. They showed that pressure could restore the luminosity of luminous bacteria exposed to an anaesthetic agent and conducted a similar experiment on tadpoles treated with ethanol. Tadpoles exposed to 2.5% ethanol ceased swimming due to narcosis. Swimming was also inhibited by pressures of 200-300 atm. However, if pressures of approximately 100 atm were applied to tadpoles treated with ethanol their swimming activity was restored (Fig. 4). The pressure reverses the effects of the ethanol which is acting as a general anaesthetic.

This phenomenon, the pressure reversal of anaesthesia, has since been observed with other animals using helium as the pressurising agent.[23,24] For the Italian great crested newt 34 atm of nitrogen acts as an anesthetic and reduces the ability of the animals to right themselves (Fig.5). As additional pressure is applied, the performance of the animals is restored to normal levels. Similar observations have been made using nitrous oxide as the anaesthetic. This antagonism has been observed in mammals, where mice have remained active at 280 atm at concentrations of anesthetic that would normally be lethal. Since

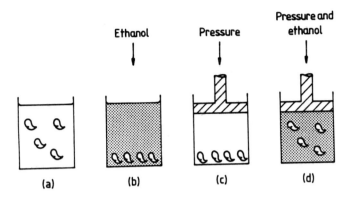

4. The effects of pressure and ethanol on the swimming activity
of tadpoles.

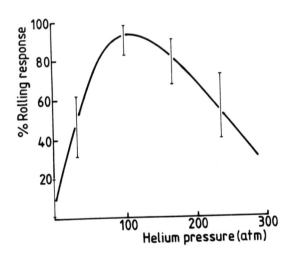

5. The rolling response (ability to follow the slow rotation of
their container) of newts exposed to 34 atm of nitrogen
in the presence of additional pressures of helium.   The
nitrogen acts as an anaesthetic and its effect is reversed by
increasing pressure.

280 atm is about twice the normal lethal pressure for mice we see
that not only does pressure reverse the effects of anaesthetics
but that anaesthetics also ameliorate the effects of pressure.

One hypothesis that seeks to account for this antagonism
assumes that the concentration of molecules at the active site is
the critical factor determining anaesthetic action. Thus it is
natural to suppose that pressure reverses anaesthesia by
'squeezing out' the dissolved molecules, and calculations show
that between 300 and 1000 atm would be required to reverse
anaesthesia. However, these values are too high when compared with
the observed value of approximately 100 atm and the squeezing-out
effect cannot be regarded as an adequate theory to account for the
pressure reversal of anaesthesia.

An alternative model that gives reasonably good prediction
of the antagonism is the so-called critical volume hypothesis
(Fig.6)[24]. This assumes that when the anaesthetic gas dissolves
at the (undefined) hydrophobic site of action of general
anaesthetics, it produces a fractional expansion given by:

$$\Delta V/V = \gamma_a P_a V_a$$

where γ_a, P_a and V_a are the solubility coefficient, the partial
pressure and the partial molar volume of the anaesthetic gas. The
application of hydrostatic pressure will lead to a reduction in
volume by:

$$\Delta V/V = -\beta P_T$$

where β is the isothermal compressibility and P_T the total

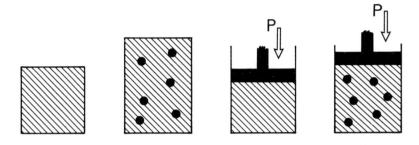

6. Diagramatic representation of the constant volume hypothesis.

pressure. We suppose that the function of the site is unimpaired
if the effects of the anaesthetic and the pressure are such as to
maintain the site at its original volume.   Then:

$$\gamma_a P_a V_a = \beta P_T$$

This equation enables us to calculate the conditions under which
pressure reversal of anaesthesia might be expected to occur. Since
the physical chemical evidence points to a hydrophobic site,
γ_a, β and V_a may be estimated with some precision.   The
converse aspect of this antagonism - the amelioration of the
effects of pressure by anaesthetic gases - can also be estimated
in the basis of this model and it has been used to estimate the
optimum mixtures for divers to breathe. Nitrogen (an anaesthetic
gas at high partial pressures) is often added to the oxy-helium
mixtures used in deep diving (tri-mix diving) to ameliorate   the
effects of pressure.   We will return later to the practical
applications of tri-mix diving.

Mechanism of the Action of Pressure

In recent years much effort has been put into elucidating
the neurophysiological changes induced by high pressures. Much
of this work has of necessity been indirect, with pharmacological
investigations playing the major part. The observation of HPNS in
the lower vertebrate species and in immature mammals which have
yet to develop cortical activity suggested that its symptoms
arise from effects on the sub-cortical regions of the central
nervous system[25]. Neuroanatomical studies have confirmed this view
that pressure has a subcortical site of action[26,27]. The
neuroanatomical studies have also indicated that pressure has no
direct action on the cortex, but it is now recognised that the
cortex provides a substantial measure of inhibition to counter the
convulsive activity that arises from subcortical regions. The
observation that reserpine, a monoamine depleting alkaloid,
abolishes the compression-rate dependence of HPNS (and is
therefore most effective in lowering convulsion onset-pressures
during slow compressions)[25] suggests that the cortical control
mechanism may account for the compression-rate dependence of HPNS
and the remission of signs that frequently occurs on prolonged
exposure to pressure. The descending inhibition originating from
the cortex appears to be primarily dependent on noradrenaline.[27]

The subcortical action of pressure suggested an analogy with
the convulsants picrotoxin and strychnine. Picrotoxin is
thought to antagonise the actions of the inhibitory
neurotransmitter GABA at subcortical sites. It was observed that
some agents that potentiate the effects of GABA postpone the onset
of high-pressure seizures. However, the discovery that

muscimol,[28] a GABA agonist, and baclofen,[26] a GABA analogue, do not confer protection against HPNS indicates that GABA may not play a direct role in determining HPNS. More recently it has been shown that a number of more specific GABA-enhancing agents do not prevent the behavioural changes associated with high pressure.[29] Experiments with the rat cervical ganglion have shown that pressure does not affect the GABA-sensitive response of this preparation.[30]

Although these results indicate that some drugs which influence the actions of GABA and other neurotransmitters can effect the onset of HPNS,[31] there is at present no evidence that these transmitters mediate the primary effect of pressure. However, in the case of glycine there is more reason to suppose that it plays a major role in determining the effects of pressure on the central nervous system. In this context the centrally acting muscle relaxants related to mephenesin have proved particularly interesting. We found that[26]

1. The aromatic propandiols mephensin and methocarbamol give excellent protection against seizures caused by pressure. They are also effective against seizures induced by strychnine.

2. The aliphatic analogues meprobamate and carisoprodol fail to protect against the convulsions associated with HPNS or those due to strychnine.

3. The relative potencies of the ortho, meta, and para isomers of mephensin are the same for protection against both pressure and strychnine - induced convulsions.

4. This correlation of the effectiveness of drugs acting against pressure and strychnine is not matched with other convulsants such as metrazol (Table 2).

Table 2   Action of Anticonvulsant Drugs[a]

Drug	Protection against seizures induced by			
	Pressure	Strychnine	Picrotoxin	Metrazol
Phenobarbitone	↑ ↑	↑ ↑	↑ ↑	-
Valproate	↑ ↑	↑ ↑	↑ ↑	↑ ↑
Diphenylhydantoin	0	0	0	0
Ketamine	↑ ↑	↑ ↑	-	-
Althesin	↑ ↑	↑ ↑	-	-
Mephenesin	↑ ↑	↑ ↑	0	0
Methocarbamol	↑ ↑	↑ ↑	-	0
Meprobamate	0	0	-	↑ ↑
Carisoprodol	0	0	-	↑ ↑
Baclofen	0	0	-	↑ ↑
Diazepam	↑	↑	↑ ↑	↑ ↑
Flurazepam	↑	-	↑ ↑	-

a ↑ ↑, Good protection; ↑, moderate protection; 0, little effect; -, no information available.

The author is indebted to Dr. F. Bowser-Riley for this assessment based largely on published results.

Strychnine is believed to act postsynaptically on glycinergic inhibitory neurons, and the pharmacological evidence outlined above suggests that pressure acts in a similar manner.

Further evidence for this point of view can be obtained from the dose-response curves for strychnine measured at high pressures of helium.[32] It was found that the effects of strychnine and pressure in inducing seizures were strictly additive, thus suggesting that both agents may share a common mechanism in the production of convulsions. In contrast, with picrotoxin which is thought to antagonise GABA, highly non-additive behaviour was observed[32] (Fig. 7). Figure 7 also illustrates the effects of simultaneous administration of strychnine and pressure in the presence of mephenesin. The effects of both strychnine and pressure are ameliorated to the same degree by mephenesin, so that the effects of the two agents remain additive, although higher doses are required in the presence of mephenesin to obtain the original level of response. The actions of strychnine and pressure are both characterized by extremely steep dose-response curves for which ED_{33}/ED_{67} = 1.10 $\pm$ 0.05. It is interesting to note that dose-response curves of this unusual degree of steepness are also observed with general anaesthetics.

The fact that glycine is not thought to act as a neurotransmitter in certain invertebrates[33] provides the possibility of investigating its involvement in the effects of pressure more directly. In particular, we have sought to elucidate its role in the pressure reversal of anaesthesia. A comparative study was carried out on the effects of pressure and anaesthetics on the electrically stimulated swimming activity of freshwater shrimps (<u>Gammarus pulex</u>) and tadpoles (larvae of <u>Rana</u>

<u>temporaria</u>) under similar experimental conditions.[34]   The measure
of swimming activity was the fraction of the total population not
resting on the bottom when this region was exposed to an
electrical stimulus. The anaesthetics chloroform, diethyl ether,
ehtanol, halothane, and sodium pentabarbitone were employed, and
their potencies were found to be very similar for both species.
However, the effect of pressure was strikingly different for
shrimps and tadpoles.   For the tadpoles, pressure clearly

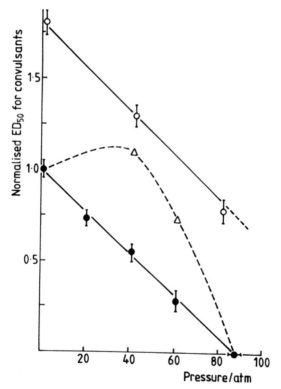

7. Effect of pressure on the doses of strychnine (●) and
picrotoxin (Δ) required to produce seizures in mice.
The upper line (O) shows the results obtained for strychnine
after the administration (130 mg kg^{-1} s.c.) of the
anticonvulsant drug mephenesin.

reversed the effects of the anaesthetics and enhanced the swimming
activity (Fig. 8). In contrast, for the shrimps, no pressure
reversal was observed and pressure enhanced the inhibitory effects
of the anaesthetics (Fig. 9). In light of our previous
investigations on the action of pressure, it is possible that the
lack of pressure reversal may arise from the absence of glycine as
a neurotransmitter in the crustacean central nervous system (CNS).

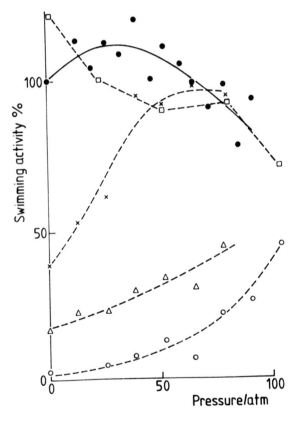

8. Swimming activity of tadpoles (larvae of <u>Rana</u> <u>temporia</u>) in the
presence of ethanol: ● no ethanol, □ 0.01 M ethanol, x 0.20 M
ethanol, Δ 0.265 M ethanol, O 0.275 M ethanol.

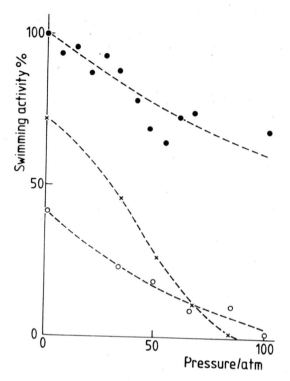

9. Swimming activity of shrimps (<u>Gammarus pulex</u>) in the presence
of ethanol: ● no ethanol, x 0.15 M ethanol, O 0.25 M ethanol.

Whether the action of anaesthetics is also mediated by
glycine so that the antagonism of anaesthetics and pressure
results from opposing perturbations of the same process or whether
the antagonism is purely physiological has yet to be established.

The design of drugs to protect against the effects of pressure

The observation that strychnine and pressure may share a
common mechanism, involving the glycine receptor protein, by which
they produce convulsions has been used in the design of drugs that

might protect divers against the effects of high pressures. Inspection of the structures of strychnine and a number of other glycine antagonists[35] suggests that they share as a common structural feature a benzene ring linked to an oxygen or nitrogen atom with a positively charged nitrogen atom some 4.5 A distant. For strychnine (Figs. 10, 11) computer modelling indicates that the N-N distance is 4.65 A and that the two nitrogen atoms and the benzene ring lie in the same plane. It is probable that the nitrogen of the strychnine molecule, which is positively charged at body pH, binds to the site that would normally be occupied by the nitrogen of the glycine molecule (Fig. 12).

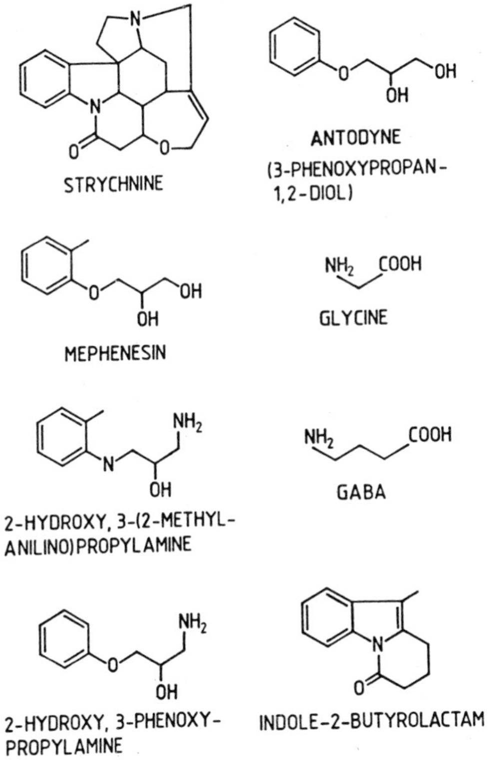

STRYCHNINE

ANTODYNE
(3-PHENOXYPROPAN-
1,2-DIOL)

MEPHENESIN

GLYCINE

2-HYDROXY, 3-(2-METHYL-
ANILINO)PROPYLAMINE

GABA

2-HYDROXY, 3-PHENOXY-
PROPYLAMINE

INDOLE-2-BUTYROLACTAM

10. Structures of substances referred to in text.

11   12

11. Structure of strychnine with that part of the molecule
believed to confer convulsant activity emphasised.

12. Relationship of the structure of the glycine molecule to that
of strychnine.

Other simple compounds which incorporate these structural
features have been shown to act as convulsants. Berger and
Lynes[36] synthesised a number of 'dinitrogen' convulsants such as
2-hydroxy, 3(2-methylanilino) propylamine (Fig. 10) which
contained basic features common to many glycine antagonists.
Substances which replace the anilino nitrogen with an oxygen atom
such as 2-hydroxy-3 phenoxypropylamine (Fig. 10) also act as
convulsants.

In contrast, compounds in which the positive nitrogen is
replaced by an oxygen, such as antodyne or mephenesin (Fig. 10),
prove effective antagonists of both strychnine and pressure.
Computer modelling indicates that one of the most energetically
stable configurations of mephenesin matches closely that of
strychnine but leaves the region of the positive nitrogen clear,
thus enabling glycine to bind (Fig. 13a). In this confirmation
mephenesin has its secondary hydroxyl group near the probable

binding site of the carboxylate moiety in glycine and may thus promote glycine binding. A slightly higher energy conformation of mephenesin would place an oxygen atom close to the site of the positively charged nitrogen which could also enhance glycine binding (Fig. 13b). We have used these models of structure-activity relationships as a guide when synthesising a range of some forty compounds capable of modifying the action of pressure on the central nervous system[37]. The results of tests on mice confirmed that any compound that placed a positive nitrogen atom approximately the same distance and direction from a benzene ring as occurs in strychnine potentiated the effects of pressure. Conversely compounds that placed a negatively charged group in this location were found to protect against the effects of

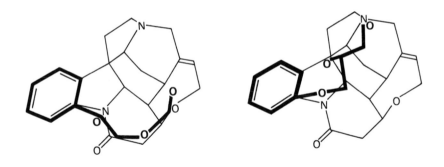

13. Two possible fits of the mephenesin molecule to that of strychnine. The diagram on the left represents the lowest energy conformation.

pressure as did compounds that were capable of hydrogen bonding to the carboxylate group of glycine. Compounds that mimic much of the structure of strychnine but do not possess groups capable of interacting with the glycine binding site might be expected to antagonise the effects of strychnine without conferring protection against the effects of pressure. An example of this behaviour was found with indole-2-butryolactam (Fig. 10). This serves to confirm the view that pressure acts by causing a conformational change in the glycine receptor protein which inhibits glycine binding.

To develop new drugs to combat the effects of pressure we hope to combine the binding but non-convulsant regions of the strychnine molecule with the structurally active features of the pressure antagonists related to antodyne. This approach will, we hope, produce powerful strychnine antagonists and drugs that will find practical application in protecting divers against the adverse effects of high pressure.

Conclusion

In recent years divers in both the United States and Great Britain using Tri-mix (nitrogen added to the helium-oxygen mixture to take advantage of the anaesthetic-pressure antagonism) have reached depths of over 2,200 ft. in simulated dives. Dives to such depths are difficult and expensive operations; it can take a week to reach depth and as much as two weeks to decompress. Nevertheless many divers have survived at depths over 2,000 ft. The scale of this achievement can be put in context by remembering that in the relatively recent past the British naval diving limit was set at 185 ft. This great advance in diving capability is to a considerable extent due to the efforts of chemists. They have helped in the clarification of the mechanisms involved in narcosis and the effects of pressure. There is every reason to suppose that the application of physico-chemical principles will lead to further advances and perhaps to the discovery of a fourth and even more unexpected surprise that will enable divers to proceed to ever greater depths.

Acknowledgements

Much of the work discussed in this lecture has been carried out by past members and current members (Dr. S. Daniels, Dr. F. Bowser-Riley, Mr. W.A.G. Hill and Mr. D. Price) of the Oxford Hyperbaric Group supported by grants from the Wellcome Trust, the Medical Research Council and the U.S. office of Naval Research. The syntheses described in the final section were carried out in collaboration with Dr. E.W. Gill and assistance with the computer modelling was provided by Dr. W.G. Richards, Mr. D. Ricketts and Mr. A. Novaks.

The author is indebted to Professors W.D.M. Paton and J.S. Rowlinson and Mrs. R.A. Smith for their comments on the manuscript.

1.   J. Priestley, *Phil. Trans*. Roy. *Soc.* 1776, **66**, 226.

2.   J. Priestley, *Phil. Trans*. Roy. *Soc.* 1772, **62**, 147.

3.   K.W. Miller and E.B. Smith in 'A Guide to Molecular Pharmacology and Toxicology', R.M. Featherstone (Ed), New York, Dekker, 1973.

4.   J. Ferguson, *Proc. Roy. Soc. London* 1939, **127B**, 387.

5.   L. Pauling, *Science* 1961, **134**, 15.

6.   S.L. Miller, Proc. Nat. Acad. Sci. 1961, 47, 1515.

7.   K.W. Miller, W.D.M. Paton and E.B. Smith, Nature (London)
     1965, 206, 574.

8.   R.D. Miller et al., Anaesthesiology 1969, 31, 301.
     S.C. Cullen et al., ibid. 1969, 31, 305.

9.   H.H. Meyer, Arch. Exp. Pathol. Pharm. 1899, 42, 109.

10. E. Overton, "Studien uber die Narkose" Jena', G. Fisher,
     1901.

11. K.H. Meyer, Trans. Faraday Soc. 1937, 33, 1062.

12. K.W. Miller, W.D.M. Paton, E.B. Smith and R.A. Smith,
     Anaesthesiology 1972, 36, 339.

13. L.J. Mullins, Chem. Rev. 1954, 54, 289.

14. F.H. Johnson, H. Eyring and M.J. Polissar, 'Kinetic Basis of
     Molecular Biology', New York, Wiley, 1954.

15. J. Hale, R. Keegan, E.B. Smith and T.J. Snape, Biochim.
     Biophys. Acta, 1972, 288, 107.

16. A.J. Middleton and E.B. Smith, Proc. Roy. Soc. Lond. 1976,
     193B, 173.

17. N. Franks and W.R. Lieb, Nature (London) 1984, 310, 599.

18. P. Bennett, Proceedings of 3rd Underwater Physiology Symp.,
     C.J. Lambertsen (Ed), Baltimore, Williams and Williams, 1967.

19. R.W. Brauer et al., Fed. Proc. Am. Soc. Expt. Biol. 1968, 25,
     202.

20. P. Regnard, 'Recherches experimentales sur les conditions
     physiques de la vie dans les eaux', Paris, Masson, 1891.

21. K.W. Miller, W.D.M. Paton, W.B. Streett and E.B. Smith, Science 1967, 157, 97.

22. F.H. Johnson and E.A. Flagler, Science 1951, 112, 91.

23. M.J. Lever, K.W. Miller, W.D.M. Paton and E.B. Smith, Nature London 1971, 231, 368.

24. K.W. Miller, W.D.M. Paton, R.A. Small and E.B. Smith, Molec. Pharm. 1973, 9, 131.

25. R.W. Brauer, R.W. Beaver and M.E. Sheenan, Proc. 6th Symp. Underwater Physiol., (Ed Shilling and Bennett), Bethesda, FASEB, 1978, p.48.

26. F. Bowser-Riley, Phil. Trans. Roy. Soc. 1984, 304B, 31.

27. F. Bowser-Riley, W.D.M. Paton and E.B. Smith, Br. J. Pharm. 1981, 74, 820.

28. A.R. Bichard, H.J. Little and W.D.M. Paton, Br. J. Pharm. 1984, 74, 221.

29. M.J. Halsey, J.-C. Rostain and B. Wardley-Smith, J. Physiol. 1984, 350, 25.

30. H.J. Little, Br. J. Pharm. 1982, 77, 209.

31. A. Angel, M.J. Halsey, H. Little, B.S. Meldrum, J.A.S. Ross, J.-C. Rostain and B. Wardley-Smith, Phil. Trans. Roy. Soc. 1984, 304B, 85.

32. F. Bowser-Riley, G. Tyres, W.D.M. Paton and E.B. Smith, J. Physiol. 1984, 357, 19.

33. H.M. Gerschenfeld, Physiol. Rev. 1975, 53, 1.

34. E.B. Smith, F. Bowser-Riley, S. Daniels, I.T. Dunbar, C.B. Harrison and W.D.M. Paton, Nature (London) 1984, 311, 56.

35. D.R. Curtis and G.A.R. Johnston, <u>Ergebn. Physiol</u>. 1974, <u>69</u>,
     98.

36. F.M. Berger and T.E. Lynes, <u>J. Pharm. Exp. Therap</u>. 1953, <u>109</u>,
     407.

37. F. Bowser-Riley, W.A.G. Hill and E.B. Smith, <u>J. Physiol</u>.
     1986, in press.

Recovery and Purification of Industrial Gases Using Prism® Separators

By I. W. Backhouse

MONSANTO EUROPE SA, AVENUE DE TERVUREN 270, I I 50 BRUSSELS, BELGIUM

Historical Introduction

Monsanto, whose heritage is founded in chemicals, ventured into the development of membrane separation systems in the early 1960's. At that time research was directed into liquid separations based on flatform cellulose acetate. By 1967 the effort was well established with some U.S. Government contracts sponsoring research into desalination. Monsanto was also working on various possibilities to improve a number of processes operated within the company and at about this time membranes were started to be produced in the form of hollow fibres.

1974 saw the decision taken to redirect research from liquid into gas separations, where the opportunities appeared to be greater. In particular, those involving hydrogen looked promising, especially in view of the four-fold increase in the price of crude oil and thus hydrogen. At about the same time polysulphone was identified as the preferred polymer.

By 1975 the first pilot scale membrane module was being tested at the then Monsanto Texas City plant on hydrogen and carbon monoxide separation. The following year the first full scale Prism® separator was ready for industrial trials. During 1977 the first system of Prism® separators was installed at Texas City to debottleneck a cold box and produce oxo alcohol synthesis gas by first intent.

This was followed closely in 1978 by a hydrogen recovery unit at Monsanto Pensacola and an ammonia plant purge gas recovery unit at Monsanto Luling in 1979 (Fig 1) . Thus, with substantial operating experience gained in

265

FIGURE 1

house, Monsanto commercially launched Prism® separators in November 1979.
Prism® separators were intially targetted for hydrogen separations such
as could be found the ammonia, oil refinery and petrochemical industries.
Since that initial launch the range of specific separations has expanded
and now also covers carbon dioxide/hydrocarbon and oxygen/nitrogen
separations.

Seven years later over 100 systems, utilising more than 1000 Prism®
separators, are operating in more than 20 countries worldwide. This,
perhaps more than anything else, indicates that membranes for gas
separations have come of age and, as a unit operation, are here to stay.

The Prism® Separator

Assuming Fickian diffusion of a dissolved gas in an isotropic dense
membrane the following equation can be derived for each component.

$$Q \quad = \quad \frac{P * A * (pf - pp)}{L}$$

where   Q   rate of flux

P   permeability coefficient

A   membrane area

pf   feed partial pressure

pp   permeate partial pressure

Transport of molecules across the membrane occurs by solution of the
molecule at the high partial pressure surface, diffusion across the
membrane under the pressure gradient and desorption at the low partial
pressure surface. The permeability coefficient for a particular
component can therefore be described as the product of the solubility
and diffusion coefficients.

As indicated in the historical introduction, the semi permeable membrane
utilised in the Prism® separator is based on polysulphone formed as
hollow fibres. In fact, generically speaking, it is an assymetric,
composite and semi permeable membrane formed as a hollow fibre.

The membrane is asymmetric because, as all other membranes that are formed
from solution, there is a thin (0.1 to 1 % of total thickness) outer
layer of dense polymer supported by porous substructure (Fig. 2).
Unfortunately it is almost impossible to produce a membrane with an ultra
thin dense layer necessary for high gas flux rates, but without
imperfections. These imperfections take the form of surface porosity and
a porosity of as small as 0.000001 can render the membrane useless. These
imperfections can be overcome by use of a coating of a material of
relatively high permeability, but low selectivity, to plug the surface
porosity and thus forming a composite membrane.

Forming the membrane as hollow fibres has two distinct advantages over the
alternate of flatform be it as stacks or spiral wound. The primary
advantage of the hollow fibre is its inherent capability to withstand high
differential pressure and to provide an excellent self-supporting
structure. Secondly the smaller the hollow fibre the higher the surface
area to volume ratio which is an important factor in determining the ultimate
size of a gas separation system. The size of the fibres relative to a paper
clip can be seen in Figure 3. The largest disadvantage of hollow fibres is
that at high gas fluxes small bore diameters can result in high bore
pressure drop which affects gas separation efficiency. This problem can
be minimised by fabricating the hollow fibre with larger bore diameters.

The choice of polysulphone for fabrication of a gas separation membrane was
based on several factors. All polymers exhibit gas permeation properties,
but clearly the ideal membrane would have both high permeability and high
selectivity. However it is in the nature of most common polymers that these
two properties never go hand in hand. So the choice of a polymer must weigh
the relative advantages of these two properties. Polysulphone, with both
good selectivity and permeability, fits the description. For any membrane

FIGURE 2

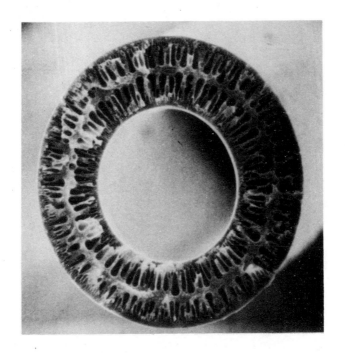

FIGURE 3

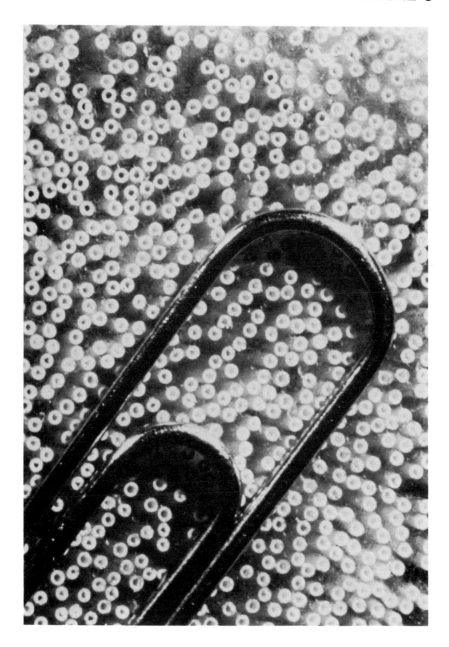

to be used in industrial applications it must able to withstand the
presence of commonly occurring compounds such as water and hydrocarbons.
Polysulphone exhibits excellent environmental resistance to a wide range of
chemical compounds providing that relative saturation does not exceed 100%.
Polysulphone also possesses very good mechanical properties enabling
self-supporting hollow fibres to be fabricated. These properties are
stable up to 100 deg C allowing a wide range of operating conditions.
Finally and perhaps most important polysulphone is a generally available
bulk polymer thereby ensuring better quality control of the final product
and at the lowest possible price.

Even the most ideal membrane for a specific gas separation is worthless
unless it can be fashioned into an efficient separator. Gases must be
conducted to and from the membrane efficiently to fully utilise the
intrinsic transport properties and the available driving force for
separation. A well designed separator ideally has a uniform velocity over
the entire membrane surface but this ideal is impractical if not
impossible. Design arrangements must always accommodate feed and product
entrances and exits which introduce stagnation points and obstructions to
flow. Changes in volumetric flow through the separator also affect local
velocities. Pressure drop to induce flow on both sides of the membrane is
equally important. Low pressure drops improve the available partial
pressure driving force but lead to non-uniform channel velocities. Thus any
membrane gas separator must pay important attention to velocity in order to
achieve an efficient separator.

Monsanto's Prism® separator design consists of a compact bundle of
thousands of hollow fibres encapsulated in a high pressure shell similar in
arrangement to a shell and tube heat exchanger (Fig. 4). The fibre bundle
is sealed at one end and encased in a tube sheet at the other and when
mounted vertically the bundle fills the shell ensuring uniform gas
distribution. Feed gas enters at the base of the shell side of the fibre
bundle and exits at the top. As the gas passes over the membrane, faster
permeating gases preferentially migrate into the fibre bore and exit via
the base of the fibre bundle. This then produces one stream lean in the

FIGURE 4

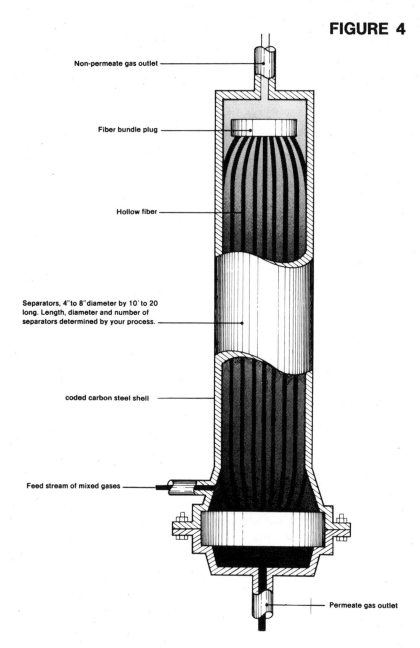

Non-permeate gas outlet

Fiber bundle plug

Hollow fiber

Separators, 4"to 8"diameter by 10' to 20 long. Length, diameter and number of separators determined by your process.

coded carbon steel shell

Feed stream of mixed gases

Permeate gas outlet

faster permeating gas at essentially the feed pressure. Typically the
pressure drop per separator on the shell side is less than 0.2 bar. The
other stream is rich in the faster permeating gas at a pressure of 10 to
50% of the feed pressure. Several separators may be connected in series
or parallel, depending on velocity and pressure drop constraints, in order
to achieve the requirements of a particular application.

At this juncture it is perhaps noteworthy to state that, since all
components will permeate to some extent, an ideal separation, producing one
or more ultra-pure streams, can not be achieved. In situations where
impurity level does not exceed 50% and impurity removal does not exceed one
order of magnitude e.g. 50% to 95% or 90% to 99% then efficient bulk
separations can be readily attained using a single stage system. Impurity
removal exceeding one order of magnitude can be achieved by recycling
and/or cascading which is equivalent to reboiling and/or refluxing in a
distillation column. However these improvements, as in distillation, can
only be achieved at the expense of more energy and/or capital.

Membrane Performance Characteristics

Generally the thickness of the membrane is indeterminate and variable so
that design of systems depends entirely on empirical determination of the
coefficient P/L. With a computer and an adequate geometrical description of
the separator a rigorous, multi-component analysis can yield specific
membrane permeability coefficients which will predict performance that is
in agreement with measured data. This approach results in very accurate
performance predictions but, unfortunately, locks the unit operations
design in computer software as no ready formulas are available to describe
a separator operating in a multi-component gas service.

The main variables affecting membrane performance are temperature of
operation, feed gas flowrate, membrane area, feed gas composition and
pressures of the feed and permeate.

Gas permeability follows an Arrhenius type of equation in that it is proportional to the exponent of the negative reciprocal of the absolute temperature. As well as increasing permeability, increasing temperature reduces the relative saturation of components in the feed gas thereby reducing pretreatment requirements. However these benefits must be traded off against lower membrane mechanical strength and lower selectivity. Thus for each application there is an optimum operating temperature range. Generally this variable is left as a fine tuning device usually to be used during the early days of operation of a system.

Feed gas flowrate and membrane area are linked together in that it is the ratio of these two that determine changes in performance. For example at high ratios the residence time is low which results in low recovery of the faster permeating gases but at high purity. As this ratio decreases and residence time increases more of the faster and some of the slower gases permeate thereby increasing recovery at the expense of a slightly reduced purity. At low ratios and long residence times the recovery is very high and purity lower than achieved at low to moderate recoveries.

The effect of feed composition on a specific performance for a given flowrate to area ratio is that as the concentration of the faster gas increases both the product purity and recovery increase. As the concentration decreases both the purity and recovery of the product decrease although the former to a much lesser extent than the latter.

Increasing feed pressure for a given flowrate and product recovery both reduces membrane area requirements and increases product purity. Increasing permeate pressure for a specific flowrate to area ratio reduces recovery and to a lesser extent the product purity.

The permeability coefficient is a function of molecular size, molecular shape, polarity and chemical affinity. Table 1 below shows the spectrum of polysulphone gas permeabilities for some common gas molecules.

Table 1

Relative Permeation Rates

H2O H2 He	H2S CO2 O2	Ar CO N2 CH4
"FAST"		"SLOW"

As might be expected small molecules such as hydrogen and helium are "fast" whereas larger molecules such as methane and argon are "slow". Anomalies such as carbon dioxide and water, which are "faster" than would be expected for their molecular size, have high solubilities in polysulphone which, as will be recalled, directly affects their permeability coefficient.

Based on the separating characteristics of polysulphone hydrogen can be separated from nitrogen, methane and argon such as occurs in the high pressure purge stream from an ammonia plant. Hydrogen can be separated from a mixture of hydrocarbons as occurs in many oil refinery purge streams and also separated from carbon monoxide which is frequently required on a petrochemical plant.

Due to the higher permeability of carbon dioxide relative to methane and higher hydrocarbons purification of the hydrocarbons or bulk removal of the carbon dioxide can be easily achieved. Purification of hydrocarbons from carbon dioxide is required for such applications as upgrading biogas or conditioning of natural gas streams. Recovery of carbon dioxide from hydrocarbons is encountered particularly in enhanced oil recovery where carbon dioxide is used to reduce oil viscosity in the oil well.

Oxygen is appreciably faster than nitrogen thereby allowing efficient separation of air to produce inert gas. The economic nitrogen purity level that can be obtained is limited by the order of magnitude reduction in impurity which for air is about 2% oxygen. However this is usually sufficient for inerting purposes and, in situations where space and weight are at a premium, membrane technology has a major advantage.

Helium, like hydrogen, also exhibits high permeability relative to nitrogen, oxygen and argon. This property has found an application for upgrading the purity of helium used for breathing gas mixtures in deep sea diving.

The Experience

Table 2 below summarises by separation the current 113 installations of Prism® separators.

Table 2

Separation	Application	Installations	Total KNM3/HR	Largest KNM3/HR
H2/N2,CH4,Ar	Ammonia	47	415	28
H2/CnHm	Refinery	15	192	40
H2/N2,CH4,CO	Petrochemical	21	120	61
CO2/CH4,CnHm	EOR/Biogas/NG	11	257	84
O2/N2	Inert Gas	18	5	2
He/N2,O2,Ar	Diving Gas	1	1	1

As can be seen from the above table the initial predictions of 1974 have proved correct in that the majority of installations have been dedicated to hydrogen recovery. However it should be noted that the most economic separations by membrane occur when high differential partial pressures are available. Thus, in situations where the feed pressure or concentration is high and the permeate pressure or concentration is low, membrane technology is at its best. This is reflected in the impact of membrane technology in hydrogen recovery from ammonia plant purge gas where nearly 50 systems have been installed. In this particular application the gas is available at high pressure, hydrogen recovery is more important than hydrogen purity and there is a convenient low pressure point to recycle the hydrogen.

The oldest unit in operation is that located at the Texas City site, which was previously owned by Monsanto. As indicated in the introduction it has now been in operation since 1977 and is very close to completing 9 years of service. The Texas City site has a reformer producing hydrogen and carbon monoxide, the latter for acetic acid production and a combination of the two for methanol and oxo alcohol synthesis. A cold box had been installed to separate a sidestream of reformer gas into its constituent components. The oxo alcohol plant required a hydrogen to carbon monoxide ratio of 1.3 to 1. This was obtained by diluting the reformer gas with a hydrogen to carbon monoxide ratio of 3.1 to 1 with pure carbon monoxide from the cold box. Unfortunately the cold box was bottlenecked limiting production of acetic acid and the carbon monoxide produced required recompression to mix it with the reformer gas. The Prism® system was installed to treat the reformer gas directly to produce the oxo alcohol synthesis gas at the correct ratio by first intent and at the required pressure. The installation then debottlenecked the cold box and produced, as a by product, hydrogen for use on the methanol plant.

The highest pressure units in operation are those installed on ammonia plants for the recovery of hydrogen from the synthesis loop purge. Typically these units operate with feed pressures in the range of 100 to 140 bars and differential pressures of 70 to 100 bars. Ammonia plants can be broken into three major processing stages, synthesis gas production, synthesis gas compression and ammonia synthesis. Using a pressure of 25 to 30 bar natural gas or naphtha is reformed with steam and air to produce a synthesis gas containing stoichiometric quantities of hydrogen and nitrogen. After removal of carbon dioxide the synthesis gas still contains 0.5 to

1.5% methane, argon and, in some cases, helium which are inert to the
ammonia synthesis. The synthesis gas is then compressed to synthesis loop
pressure, which can range from 80 to 450 bar. Some of these inerts are lost
by solution in the product ammonia but in most cases a purge stream is also
required to maintain reactant partial pressures. As this purge stream can
represent 5 to 6% of the hydrogen manufactured and is usually used only for
its fuel value, interest in purge gas recovery is high. Prism® systems
can conveniently recover over 90% of the hydrogen and 99% of the ammonia
present in the purge stream enabling either capacity to be increased or
energy saved. The driving force force for separation is the differential
pressure between the synthesis loop and the various suction pressures of
the synthesis gas compressor.

The oldest refinery unit is installed to produce hydrogen of greater than
98% purity to supplement supply from a hydrogen plant to a hydrocracker.
Source of hydrogen for this unit is the purge from a naphtha
hydrodesulphuriser unit which contains about 80% by volume and is at a
pressure of 44 bar. Recovered hydrogen is conveniently fed to the suction
of the make up compressors at a pressure of 17 bar. This unit was installed
in 1980 and has given trouble free operation during the last 6 years.

The largest system in operation is that treating nearly 85000 nm^3/hr of
casinghead gas from an oil well undergoing enhanced oil recovery based on
carbon dioxide injection . Prism® separators recover carbon dioxide that
dilutes the hydrocarbons in the associated gas after carbon dioxide
breakthrough. This system typically recovers 90 to 95% of the carbon
dioxide at purities ranging from 90 to 97%. The recovered carbon dioxide is
then combined with make up carbon dioxide for reinjection into the well.
Generally in this type of application the most economic combination of
unit operations consists of casinghead gas compression to 20 bar followed
by bulk membrane separation removing 85 to 95% of the carbon dioxide. Final
gas clean up from 15 to 20% carbon dioxide to pipeline specification is
then performed by conventional liquid absorption. Prism® separators are
ideally suited for this type of application as the flowrate and composition
of casinghead gas vary throughout the life of the project. This variation

either requires the user to install equipment to treat the maximal conditions or to be capable of expansion.

One of the most remote units is that which is installed in Wyoming and is located more than 100 kilometres from the operator. This unit removes small quantities of carbon dioxide from natural gas in order to achieve pipeline specification. Membrane technology was selected over other technologies for this application because of its ease of operation, low maintenance and ability to accept fluctuating flowrates and compositions without operator intervention. Waste permeate gas from the Prism® system is used to power the gas compressor required to boost the natural gas to pipeline pressure.

The smallest units are located on North Sea oil rigs where they are subject to severe meteorological conditions. Using compressed air at 40 bar these small units produce 10 to 1500 Nm^3/hr of inert gas containing 1 to 5% oxygen and with dew points well below 0 deg C. These units are ideal for this type of application because weight for weight they are lighter than competitive technologies and can also be "shoe horned" into restricted spaces. They have an operating advantage in that being based on a truly continuous process, maintenance and resulting downtime are minimal.

In Conclusion

The advantages of membrane technology for the recovery and purification of industrial gases can be summarised as follows.

Prism® separators typically tolerate most contaminants at levels up to several hundred ppm in the process stream and in some cases even much higher levels. Thus the necessity for pretreatment is reduced, if not eliminated, in most instances. This contrasts with other technologies that require much more stringent pretreatment.

Unlike most other gas separation technologies variations in both feed flowrate and composition to a Prism® system result in only minor changes in product flowrate and purity thus avoiding the need for operator

attention. Other technologies require operator or computer intervention to
adjust operating conditions to account for variations in both feed flow and
composition.

As the technology is based on a truly continuous process utilising no
moving parts, maintenance is low and on stream time is high. In the event
that maintenance or isolation is required, shut down and start up can be
achieved in minutes which compares favourably with other technologies.

The units are mechanically simple and compact ensuring that Prism®
separators can be retrofitted in many existing facilities. Systems are skid
mounted so that site installation is simple and rapid requiring only
bolting down, connection of tie-ins and connected and instrument hook-up.
As the units are of a modular nature plant expansions can be easily
achieved at a later date by simply adding more separators.

Prism® separators offer a simple solution to gas separation problems. One
that is so easy to operate and maintain that one operator commented that
using a Prism® separator was " like operating a piece of pipe ". They
have been commercially proven for nearly nine years in the chemical
processing and hydrocarbon refining industry and have justly taken their
place in chemical engineering as an accepted unit operation.

The Use of Membranes in the Japanese 'C₁ Chemistry Programme'

By T. Hakuta*, K. Haraya, K. Obata, Y. Shindo, N. Ito, and H. Yoshitome

NATIONAL CHEMICAL LABORATORY FOR INDUSTRY, TSUKUBA, IBARAKI 305, JAPAN

Introduction

In Japan, the petrochemical industry uses mostly naphtha to synthesize basic chemicals. However, the cost of petroleum is unstable and its supply in quantity will be tight; then the Japanese petrochemical companies will need to switch their feedstock from naphtha to coal, LNG and biomass in near future.

Considering this situation, the Japanese government adopted "C₁ Chemistry" as the National Research and Development Programme (commonly known as the Large-Scale Project) of Agency of Industrial Science and Technology (AIST) of MITI.

This programme consists of the following three themes,

(1) New technology and development of catalysts for synthesis of basic intermediates such as ethylene glycol, ethyl alcohol, acetic acid and olefins from synthesis gas.

(2) Technology of gas separation with membrane for adjustment of H_2/CO ratio to synthesize variable products.

(3) Survey of gasification processes and design of total C₁ Chemical system.

"C₁ Chemistry Programme" started from F.Y. 1980 and will be completed in December of 1986. The total budget is approximately 10.5 billion yen (68 million U.S. dollars). Fig.1 shows the detailed schedule for gas separation technology based on membrane method. The research and development indicated by lines 1 and 2 has been carried out by National Chemical Laboratory for Industry (NCLI). The others have been developed by three private companies, (1) UBE Industries, Ltd., (2) TOYOBO Co., Ltd. and (3) SUMITOMO Electric Industries, Ltd. AIST and NCLI play important roles to promote this project in cooperation with them.

281

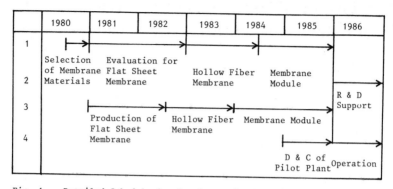

Fig. 1   Detailed Schedule for Gas Separation Technology with Membrane

The targets of membranes for gas separation in "C_1 Chemistry Programme"
were decided by AIST and NCLI before the start of this R & D, and in F.Y. 1980
five types of membranes were selected from many candidates. Tables 1 and 2
show the target values of performances for membranes and types of each gas
separation membrane, respectively.

Table 1   Targets of Gas Separation Membranes

	Pressure	Max.Temp.	Separation Factor		Permeability Coef.
	Kg/cm^2	°C	H$_2$/CO	CO/CH$_4$	cm^{3}cm/cm^{2}s cmHg
Porous Membrane Organic	>50	150	2–3		>10^{-4}
Inorganic	>50	400	2–3		>10^{-4}
Nonporous Organic Membrane	>50	100	>20	2–3	>10^{-9}

Table 2   Materials and Types of Membranes by Each Company

UBE Industries, Ltd.	Nonporous Asymmetric Polyimide Membrane, Porous Polyimide Membrane 3,3',4,4'-Biphenyltetracarboxylic Dianhydride, 4,4'-Diaminodiphenyl Ether
TOYOBO Co. Ltd.	Nonporous Asymmetric Polysulfone Membrane Bis(4-(4-aminophenoxy)phenyl)sulfone, Bis(4-hydroxyphenyl) sulfone, Isophthaloyldichloride Porous Silica Glass Membrane
SUMITOMO Electric Industries, Ltd.	Nonporous Plasma Polymerized Membrane Substrate; Porous PTFE Intermediate; Polyphenyleneoxide Active Layer; Methyltrivinylsilane etc.

Pure Gas Permeation

Permeation rates of H₂ and CO were measured for flat sheet membranes to screen available materials. Suitable polymers for gas separation membrane in "C₁ Chemistry Programme" were selected from various kinds of polyimide, polysulfone and plasma polymerized substances. Plasma polymerized membranes consisted basically of three layers and the active thin layer was prepared on the surface of silicone regine which was coated on the porous PTFE substrate by plasma. In succession, hollow fiber types of membranes were spun and those performances were measured. Most of those membranes had better performances than target values. Figs. 2 and 3 show the performances of nonporous organic membranes. Permeabilities of various gases for capillary type of porous SiO₂ glass membrane are given in Fig.4. The slope of the line in this figure nearly corresponds to the $M^{0.5}$ relationship. Permeability ratio of H₂ to CO is about 3.5 at 20°C and this value is close to the ideal value ($\sqrt{CO/H_2}=3.74$).

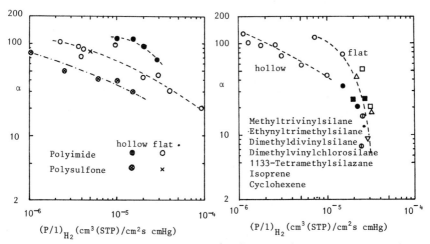

Fig. 2 Permeation Rates and Separation Factors of Polyimide and Polysulfone Membranes for H₂ and CO at 30°C

Fig. 3 Permeation Rates and Separation Factors of Various Plasma Polymerized Membranes for H₂ and CO at 30°C

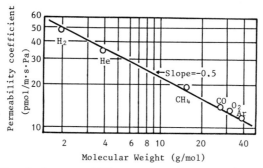

Fig. 4 log permeability vs. log molecular weight

As a typical example, Fig.5 shows the temperature dependence of permeability

of dense polyimide (PI) membrane produced by UBE Industries, Ltd. to various

gases. This membrane was 15μm thick and was prepared by casting the polymer

solution, which contained 10wt.% PI in a solvent of phenols, on a flat glass

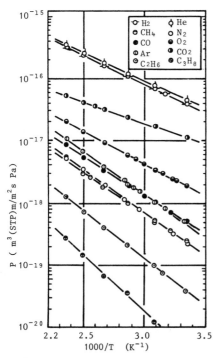

plate at 70°C and evaporating the

solvent at 100°C for 3hrs., suc-

cessively, at 300°C for 1hr. The

permeability ratios of H_2 to other

gases at 25° and 100°C are given

in Table 3.

Table 3 Relative Permeabilities
of H_2 to Various Gases of PI
Membrane at 25° and 100°C

	25°C	100°C
He	0.867	0.909
H_2	1	1
CO_2	3.47	5.12
O_2	20.5	17.5
CO	78.4	49.1
CH_4	161	87.4
N_2	163	82.9

Fig. 5 Temperature Dependence of
Permeability of Polyimide Membrane

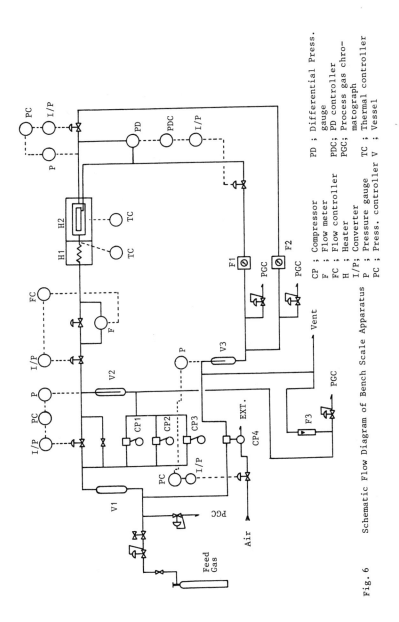

Fig. 6   Schematic Flow Diagram of Bench Scale Apparatus

Hollow fiber and capillary membranes with better performances were equipped in steel shells as modules. H_2 and CO permeabilities of membrane modules were measured by a bench scale apparatus which flow diagram was given in Fig. 6. This apparatus basically consists of four parts ; equipments for gas recirculation, process gas chromatograph for analysis of gas mixture, constant temperature bath, and temperature, pressure and flow controller. The experimental results for five kinds of membranes in F.Y.1984 and 1985 are shown in Fig. 7. H_2 permeabilities of nonporous organic membranes except plasma polymerized membrane increased but the selectivities decreased as the temperature raised. The selectivity of plasma polymerized membrane was extremely affected by the kind of polymerized monomer. The dependency of temperature on performances of porous membranes is not apparent.

Separation of H_2/CO mixtures

The degree of separation of module were measured for H_2/CO binary mixtures at various temperatures, pressures, feed compositions and flow rates. Figs. 8 and 9 show results of separation tests by polyimide and polysulfone membranes at 50°C, respectively. Hydrogen of higher concentration than 90% could be obtained from the equimolar H_2/CO gas mixture in a wide range of stage cut by permeaters with these membranes.

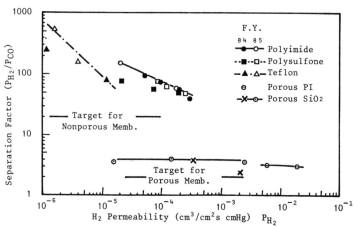

Fig. 7   Performances of Modules

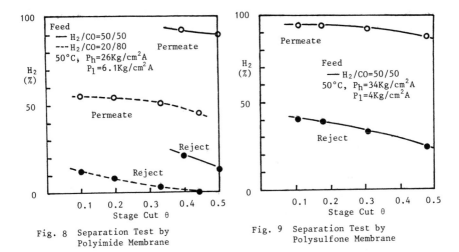

Fig. 8   Separation Test by
         Polyimide Membrane

Fig. 9   Separation Test by
         Polysulfone Membrane

A part of result in the bench scale test is also given in Table 4. In this table, α_s, α_h and α_p are calculated the following Eqs.(1),(2) and (3), respectively.

$$\alpha_s = \frac{X_p/X_p'}{X_{np}/X_{np}'} \quad (1)$$

$$\alpha_h = \frac{X_p/X_p'}{X_f/X_f'} \quad (2)$$

$$\alpha_p = \alpha_s \cdot \frac{X_{np}' - P_r X_p'}{X_{np} - P_r X_p} \cdot \frac{X_{np}}{X_{np}'} \quad (3)$$

where $X_p'=1-X_p$, $X_{np}'=1-X_{np}$, $X_f'=1-X_f$, and $P_r=P_p/P_{np}$. α_s, α_h and α_p are generally called the stage separation factor, head one and apparent ideal separation factor in perfect mixing model. P, X and X' represent pressure, compositions of more permeable light gas (H_2) and of heavy gas (CO). Subscripts f, p, np, and r mean feed, permeate, nonpermeate and ratio, respectively.

Pilot Plant Test

Pilot Plant was constructed at UBE Ammonia Co., Ltd. in Yamaguchi Pref. and has been operated from April of F.Y.1986 and the operation will be completed in December 1986. Fig. 10 shows the flow diagram of the pilot plant. Four types of membrane modules were equipped to the plant.

Table 4 Examples of H_2/CO Separation Test Results

Temp. °C	P_{np} Kg/cm²	P_p	ΔP	P_r	θ	X_f	X_p	X_{np}	α_s	α_h	α_p
1) Polyimide											
100	14.0	4.0	10	0.286	0.130	0.437	0.898	0.347	16.6	10.7	60.7
	14.0	4.0	10	0.286	0.221	0.452	0.872	0.331	13.8	8.26	52.6
	24.0	4.0	20	0.167	0.226	0.478	0.935	0.346	27.2	15.7	48.7
	44.0	4.0	40	0.091	0.353	0.474	0.909	0.285	25.1	11.1	34.9
14	40.0	12.0	28	0.317	0.111	0.461	0.964	0.400	38.6	30.0	137
25	24.9	5.0	19.9	0.232	0.265	0.466	0.959	0.313	57.1	29.8	194
2) Polysulfone											
100	23.9	4.0	19.9	0.167	0.09	0.452	0.945	0.406	25.1	20.8	40.6
	38.9	4.0	34.9	0.103	0.10	0.477	0.957	0.415	31.4	24.4	40.8
	38.9	4.0	34.9	0.103	0.40	0.471	0.901	0.227	31.0	10.2	51.7
	50.9	4.0	46.9	0.079	0.146	0.502	0.957	0.398	33.7	22.1	41.3
50	34.0	4.0	30	0.118	0.107	0.542	0.948	0.411	35.0	20.6	47.7
3) Plasma											
100	23.9	4.0	19.9	0.167	0.137	0.459	0.983	0.381	93.9	68.2	165.
	33.9	4.0	29.9	0.118	0.166	0.473	0.991	0.368	189.	123.	277.
	43.9	4.0	39.9	0.091	0.204	0.460	0.987	0.359	136.	89.1	181.
	53.9	4.0	49.9	0.074	0.098	0.475	0.997	0.421	457.	367.	554.
4) Porous SiO_2											
400	30.9	6.0	24.9	0.194	0.067	0.492	0.703	0.475	2.62	2.44	3.27
	30.9	6.0	24.9	0.194	0.177	0.493	0.697	0.452	2.79	2.37	3.55
	30.9	6.0	24.9	0.194	0.408	0.506	0.683	0.389	3.38	2.10	4.62
	45.9	6.0	39.9	0.131	0.119	0.505	0.713	0.468	2.82	2.44	3.28
	45.9	6.0	39.9	0.131	0.387	0.500	0.697	0.397	3.49	2.30	4.24
100	30.9	6.0	24.9	0.194	0.204	0.510	0.673	0.470	2.32	1.98	2.83
	30.9	6.0	24.9	0.194	0.335	0.516	0.666	0.440	2.54	1.87	3.18

Syn-gas from coal gasifier is initially cooled and removed water by drain separator. Sour gases (H_2S, CO_2, COS etc.) are absorbed in cold methanol in absorber. Feed gas after treatments is compressed and heated to operating condition, and then is supplied to the equipped modules.

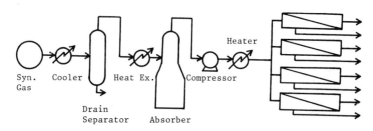

Pretreatment of Syn. Gas   Acid Gas Removal   Membrane Permeater

Fig. 10   Schematic Flow Diagram of Pilot Plant

Composition of feed gas after treatments is as follows:

H_2	47.4±3%
CO	51.8±3%
CO_2	100ppm
H_2S	2ppm
COS	2ppm
$N_2+Ar+CH_4$	3%
CH_3OH	200ppm

Operating condition is as follows;

Presssure of feed gas	$30 - 60Kg/cm^2$
Pressure of permeate	$2 - 60Kg/cm^2$
Temperature	$50 - 120°C$
Feed rate	$50 - 600Nm^3/hr/module$

Experimental Results for four types of membrane modules in pilot plant test are given in Table 5. Qs in this table are flow rate (Nm^3/hr).

Table 5   Results of Pilot Plant Test

1) Polyimide

Temp.	P_{np}	P_p	P_r	θ	X_f	Q_f	X_p	Q_p	X_{np}	α_s	α_h	α_p
95°C	29.9	11.8	0.395	0.090	44.9	147	88.0	13.3	40.6	10.7	9.0	68.5
98	30.0	6.0	0.200	0.183	45.5	197	93.6	36.1	34.7	27.5	17.5	58.5

2) Polysulfone

100	30.0	6.0	0.200	0.215	46.0	119	93.8	25.5	32.9	30.9	17.8	70.6
92	50.0	5.0	0.100	0.339	46.1	110	93.6	37.3	21.6	53.0	17.1	92.8

3) Teflone

97	30.0	5.0	0.167	0.038	46.2	150	95.1	5.7	44.3	24.4	22.6	37.5
100	50.0	5.0	0.100	0.035	46.7	150	95.6	5.2	44.9	26.7	24.8	33.7

4) Porous SiO_2

71	30.0	5.0	0.167	0.404	46.1	19.8	55.3	8.0	40.8	1.80	1.45	2.03
88	30.0	5.0	0.167	0.207	46.0	39.6	58.4	8.2	43.0	1.86	1.65	2.11

Conclusion

R & D of gas separation membranes have been carried out in " C_1 Chemistry Programme " by NCLI and private companies. Membranes developed in this programme are highly resistant to heat and chemicals that are contained in syngas as contaminants. All the nonporous organic membranes can be used steadily up to 100°C of temperature and $50Kg/cm^2$ of pressure. Porous PI and Silica glass membranes can also be used up to 150° and 400°C, respectively. These membranes clear the target values and probably have better performances than the current membranes for H recovery. Advantages in higher temperature ope-

Fourth BOC Priestley Conference

ration are as follows;

(1) The permeation rate of gas through the separation membrane is higher as operating temperature rises, so that the required membrane area is smaller.
(2) Gas separation modules can be used safely for gas of which the dew point is lower than 100°C.

The membranes developed in this programme will be used effectively not only in the adjustment of H_2/CO ratio for " C_1 Chemistry " but also in wide fields of gas separation such as H_2 recovery from an ammonia purge gas and a naphtha hydrodesulfurizer purge stream, sour gas treatments, nitrogen enrichment, CO_2 recovery in EOR, etc..

Acknowledgement

This work is a part of the results of " C_1 Chemistry Programme ", a National R & D Programme of the AIST, MITI, Japan.

The authors are grateful to UBE Industries, Ltd., TOYOBO Co., Ltd. and SUMITOMO Electric Industries, Ltd..

Applications of Membrane Technology in the Gas Industry

By B. W. Laverty and J. G. O'Hair*

BRITISH GAS, LONDON RESEARCH STATION, LONDON SW6 2AD, UK

Introduction

Membrane technology is now being used in many areas of the chemical and related industries. Within the last decade, the use of membrane technology to effect industrial gas separations has been shown to be both practical and economical. British Gas is currently examining areas in which membrane technology might be of value in the gas industry.

In this paper, we shall discuss the role which gas separation techniques play in the gas industry and where membrane technology might be most advantageously employed, and we shall conclude by briefly describing work currently being carried out by the London Research Station of British Gas to develop this technology.

Background

Gas separation processes have been at the heart of the gas industry since its inception. The first separation technique to be employed by the industry was chemical absorption. The gas-making processes of the early nineteenth century produced a crude coal gas consisting primarily of H_2, CO and CH_4, with various impurities such as CO_2, H_2S, NH_3 and N_2. The NH_3 was removed by washing with water in the gas cooling stage, which also removed any volatile tars that might have been carried over from

291

the gasification retorts. After washing, the gas was passed over slaked lime to remove the bulk of the CO_2 and H_2S. Later, in the middle of the last century, a second chemical absorption stage was introduced in which iron oxide was used to reduce the level of H_2S further still[1].

The removal of CO_2 and H_2S is a problem which has persisted in the gas industry throughout its history. Even today, these gases must be removed from some natural gases and from substitute natural gas.

The removal of these acid gases is a problem that the gas industry shares with the chemical industry in general, and as both the gas and chemical industries have advanced, new separation techniques have been developed to counter this problem. These include :

1. Improved Chemical Absorption techniques,such as the Benfield process.
2. Absorption by Physical Solvents such as methanol (the Rectisol process).
3. Direct Conversion processes such as the Stretford process.
4. Physical Adsorption processes such as pressure swing adsorption (PSA).
5. Cryogenic techniques.

These techniques for gas separation can require large, complex chemical plant and can be expensive to operate and maintain. Consequently, much chemical and chemical engineering effort is directed at improving separation techniques.

Use of Membranes for gas separations

The concept of separating gas mixtures by exploiting the relative abilities of different gases to pass through a (semi-permeable) membrane dates from the early work of Thomas Graham who reported a partial separation of gases in 1866[2]. The process by which a substance passes through a membrane is known as permeation and can be described in terms of solution of the permeant in the membrane material and diffusion within the membrane. The driving force for permeation is the difference in partial pressure across the membrane. The permeability coefficient, Q, for a gas in a particular membrane can be readily measured, and, as membranes can exhibit different permeabilities to different gases, an ideal separation factor $\alpha_B^A = Q_A/Q_B$ can be defined as the ratio of the permeability coefficients of two gases A and B.

The separation of binary mixtures under different flow regimes has been the subject of several models[3-5], and these models have been reviewed by Hwang and Kammermeyer[6]. Weller and Steiner[3] have discussed the simple case, shown in Figure 1, where a feed stream flows past a membrane to exit as a non-permeate stream at pressure P_I, and a proportion of the feed gas permeates across the membrane to form a permeate stream at a lower pressure P_{II}. The proportion of gas crossing the membrane is known as the stage cut, and is defined as $\theta = J_p/J_f$, where J_f, J_p are the feed and permeate flow-rates, respectively.

These models enable the permeate and non-permeate stream concentrations to be calculated for a specified membrane configuration and, in addition, enable the membrane area required to effect a specific degree of separation or enrichment to be calculated.

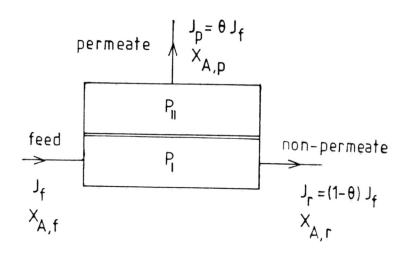

FIG.1 PERMEATION

Development of membrane Technology

Mathematical treatments of the type outlined above enabled
membrane separation systems to be designed to carry out gas
separations. However, until 1968, the only systems actually in
use were the large gaseous diffusion facilities for the isotopic
enrichment of UF_6, and for the purification of hydrogen by passage
through palladium-silver alloys[6].

The early application of membranes in gas separations was
held back by the problems of poor selectivity and/or permeabilities of early membrane materials. This meant that very large
surface areas, often in multistage configurations, were required
to provide acceptable levels of enrichment at acceptable flow

rates. Inter-stage compression in multi-stage configurations
(required to raise the pressure of a permeate stream before it
enters the second or subsequent stage of a separation unit)
increases both capital and running costs. In addition, membranes
in the form of flat sheets provided a very low membrane surface
area to permeator volume ratio, and the sheets also required
support against the high differential pressure which is the
driving force for the separation process. Provision of this
support would further increase the capital costs of installing a
membrane separation system.

These problems were common to many fields of membrane
technology and solutions developed in other areas, especially
reverse osmosis, have found application in gas separations.

The development of the asymmetric cellulose acetate membrane
by Loeb and Sourirajan[7] led to greatly improved membrane performance. By increasing the density of the skin of the membrane on
one side only, an essentially microporous membrane is rendered
non-porous on that side. Transport across the membrane is then by
molecular diffusion through this non-porous layer and thence by
Knudsen diffusion through the microporous layer. As one would
expect, these membranes show higher permeability than a physically
homogeneous membrane of the same thickness. The selectivity is
not, however, greatly reduced, and the asymmetric membrane
effectively combines the high permeability of a microporous
membrane with the high selectivity of a non-porous membrane.

The success of the asymmetric membrane led to the development
of composite membranes in which a thin layer of a chemically
different polymer is laid on top of a microporous supporting
polymer. In Type I composites, the outer polymer is much less

permeable and more selective than the support, and it is this film which is the effective permeation barrier. The outer film gives high selectivity whilst the porous support allows greater permeability than a membrane composed entirely of the selective material. In Type II composites, the outer polymer is highly permeable, and serves to fill the surface pores of the substrate preventing leaks. The substrate then affords selectivity at an acceptable permeability, whilst the outer layer prevents gas passing straight through the membrane. Because the surface area of the coating-filled pores is small, resistance to permeation is actually greater in these pores than in the substrate polymer.

Solutions to the problems of providing and supporting an adequate membrane surface area have also emerged from reverse osmosis, and are now in use for gas separation.

The spiral-wound asymmetric membrane in which a 'sandwich' comprising feed/reject channel, membrane and permeate channel is wound around a permeate off-take tube, is the basis of several commercial gas separation systems, such as those marketed by Separex, Envirogenics (Gasep), Delta Engineering (Delsep) and Grace.

Hollow fibre membranes, in which the membrane derives its support from the thickness of the walls, were used for commercial gas separations as early as 1968 by Du Pont de Nemours (Permasep) but received a boost with the introduction of the hollow fibre composite membrane which is the basis of the Monsanto (Prism) family of gas separators. Hollow fibres are also used in gas separation modules produced by Cynara.

Areas suitable for the use of membranes

The principal areas where British Gas is interested in using membrane technology can be divided into four categories :

1. Hydrogen recovery

2. CO_2 removal

3. O_2 enrichment/N_2 production from air

4. Others.

Hydrogen Recovery

British Gas was first attracted to membrane technology as a means of recovering hydrogen during the synthesis of substitute natural gas (SNG). British Gas still maintains a gas-making capacity and an active research and development programme devoted to the production of SNG from feedstocks such as light oil fractions and coal. Hydrogen is used at various stages in these processes and for reasons of economy, recovery and recycling of unreacted hydrogen is desirable.

In the British Gas Catalytic Rich Gas (CRG) process, a mixture of naphtha feedstock and steam is passed over the CRG steam-reforming catalyst to form the rich gas from which the process derives its name. Typically, rich gas might be composed of 30% CH_4 and 18% H_2, with the oxides of carbon making up the rest of the mixture. This rich gas can be reformed to make towns gas or undergo catalytic methanation which, after removal of CO_2, yields an SNG containing up to 95% CH_4.

The catalysts used in the CRG process are, however, susceptible to poisoning by sulphur-containing compounds, and the naphtha feedstock must be purified before the steam-reforming stage. This is achieved by a hydrodesulphurisation stage.

After initial start-up, the hydrogen required for this stage is sometimes supplied by recycling a portion of the rich gas. With heavier feedstocks containing the most refractory sulphur compounds, a high concentration of hydrogen is required in this recycle gas to achieve satisfactory desulphurisation. A membrane separation unit offers a much cheaper way of producing this high hydrogen concentration than the usual secondary high temperature steam reforming unit.

Even with lighter feedstocks, where upgrading of the recycle gas composition is not technically necessary, it is possible that the savings in vessel sizes, and particularly the capital and operating costs of the recycle compressor, will more than pay for the membrane separation unit.

In the case of existing plants, in certain circumstances, it might be possible to retrofit a membrane separation unit to upgrade the recycle gas to enable the plant to produce more gas or to handle heavier or higher sulphur content feedstocks than it was originally designed for.

The CRG example shows how membrane technology could be used to improve an existing process. However, in most processes, membrane technology must show itself equal, if not superior, to existing conventional gas separation technologies.

Carbon Dioxide Removal

The production of gas from oil or coal feedstocks produces large quantities of carbon dioxide arising from the need to oxidise some of the carbon in the feed to provide the energy to drive the process. The heavier the feedstock, the more carbon dioxide is formed. It is around 20% of the product gas from the CRG process and around 50% from the Slagging Gasifier/HICOM coal-based SNG process.

CO_2 also occurs in natural gas. Several of the more recently developed North Sea fields have been shown to possess appreciable amounts of CO_2, the concentration of which may need to be reduced. Since the separated CO_2 is normally vented, a separation system must minimise the amount of CH_4 lost along with the CO_2, as the extent of this 'slippage' must be taken into account when evaluating the operating costs of the system.

It had been thought that membrane technology for CO_2 removal operated most cost-effectively at high (~80%) CO_2 concentrations[8] and that conventional absorption technology would be more economic at low CO_2 concentrations. However, a recent study[9] showed that a 2-stage membrane separation system to remove CO_2 from a natural gas containing 8% CO_2 would, in fact, cost less than a conventional Diethanolamine (DEA) absorption plant. Overall capital costs for the membrane unit are shown to be approximately 25% cheaper than for the DEA plant, whilst operating costs are reduced by some 60%, in spite of the fact that hydrocarbon slippage is more than six times greater on the membrane unit. This may have been brought about by using the CO_2-rich permeate gas as a low calorific value fuel gas, thus saving on purchased energy costs.

However, in cases where hydrocarbon slippage cannot be allowed or offset against energy costs, it has been suggested that a membrane separation unit could be used in conjunction with a

conventional separation unit, e.g. a Benfield type chemical
absorption unit[10]. Use of a membrane separation unit to remove
bulk CO_2 from a gas stream, i.e. using a membrane separation unit
as a pretreatment stage, could allow a smaller, cheaper absorption
unit to carry out the final CO_2 removal. As the membrane unit
would not itself carry out a complete separation, only a small
membrane unit, capable of recovering perhaps half or two-thirds of
the CO_2 content of the gas with minimal slippage, would be
required. Costing estimates indicate that a combined system using
a membrane separator to carry out an initial, crude separation,
combined with a conventional separation system, should be more
economical than a scheme where the full separation was to be
effected by either absorption or a membrane unit alone.

Table 1

Comparison of amine and membrane processes for CO_2 removal:
(for 3,348 std. m³/h stream of natural gas containing 8% CO_2).

	Amine	Membrane
Capital cost : $	460000	340000
Daily operating costs : (as 1000 std. m³ fuel gas equivalent)	3.00	1.15
(as % of inlet hydrocarbon)	3.4	1.3

Adapted from Reference 9

An example of a possible combined membrane/conventional separation unit could be in the field of energy storage as liquefied natural gas. Prior to undergoing liquefaction, the CO_2 concentration of the feed gas must be reduced to 100 ppm or less, to prevent the formation of solid CO_2 blocking the liquefaction unit. At present, this is carried out by temperature swing adsorption (TSA) on a bed of molecular sieve. A system can be devised whereby a simple membrane system such as the one described in the previous example, could be used to remove the bulk of the CO_2 allowing the final purification to be carried out by a smaller TSA plant. Obviously, the more selective the membrane is for CO_2, the greater the role that can be allotted to membrane separation for this purpose.

An area of the gas industry where the future seems promising for the use of membrane technology is in the offshore removal of CO_2 from natural gas. Several North sea gas fields, e.g. the Sleipner field in the Norwegian sector, contain appreciable amounts of CO_2, and it is likely that future finds around the British Isles will also require removal of CO_2 (and H_2S). In addition to ensuring that the gas meets transmission standards, there are other reasons for removing CO_2 offshore at the well-head.

CO_2 in the presence of water can be corrosive, attacking pipelines. Removal of CO_2, particularly in fields where CO_2 content is high, reduces the bulk of gas which must be transported on shore, and reduces the amount of compression power required to land the gas. In cases of extremely high CO_2 content, offshore CO_2 removal could also result in reductions in the pipeline size. Bulk removal could be particularly important where Enhanced Oil Recovery (EOR) techniques are in use, as discussed below.

Membrane technology is particularly suited to offshore
applications where the area available for purification plant is
limited. Similarly, the weight of plant that can be deployed
offshore may also be limited by the constraints of platform
construction. Membrane separation units are modular in
construction, and are generally more compact than absorption
plant. They also require little maintenance and have negligible
start-up times. The problem of slippage may also be circumvented
by using the permeate gas as a low calorific value fuel gas both
for heating and to power turbines aboard the production platforms,
thus reducing energy costs. Again, as membranes with improved
separation factors are developed, membrane installations could be
further reduced in size, increasing their competitiveness and
suitability for deployment offshore.

One offshore use for membrane separation that can be expected
to be of increasing importance in the future is in the separation
of CO_2 from the associated hydrocarbon gases from oilfields which
are being exploited using EOR techniques. In the most common of
these techniques, CO_2 is injected at high pressure into oil
reservoirs to stimulate the recovery of oil. Gas is also released
from reservoirs subjected to this treatment, and arrives at the
well-head heavily diluted with CO_2, which can comprise some 80% of
the gas mixture.

Separation of CO_2 would be desirable here for several
reasons. As discussed above, pumping the gas ashore untreated
could give rise to corrosion, and would increase compression
costs. Restrictions on flaring, and renewed interest in gas
gathering pipelines might make it advantageous to recover methane
for platform heating and power and for delivery onshore. Finally,
recovery and recompression of the CO_2 for reinjection could

greatly reduce the cost of applying EOR offshore, by minimising the amount of CO_2 which would have to be purchased and transported to the production platform.

This form of EOR is presently employed onshore, particularly in the USA, and all the major manufacturers have membrane units installed in operating fields. Membrane technology is, for this application, well proven technology and is fully competitive with conventional amine absorption techniques. It has been estimated[9] that, at CO_2 levels in excess of 80%, the capital investment and operating costs of a membrane separation unit might be 50% less than those of a conventional amine absorption plant. At present, we are unaware of any membrane plants operating as part of an offshore EOR scheme, but it is believed that this will be an area of increasing importance in coming years.

Nitrogen Separation and Oxygen Enrichment

The third principal area of interest for the gas industry is in the separation of nitrogen and oxygen from air. Nitrogen is used as an inert purge gas in many processes throughout the chemical industry, and is used for this purpose in SNG production processes. Conventionally, nitrogen separation is carried out by cryogenic air separation plants, and is stored in liquid form. Few industrial plants are without either liquid nitrogen storage or nitrogen production facilities. Most polymer membranes exhibit greater permeability to oxygen than to nitrogen and, therefore, a membrane separation unit can supply nitrogen purge gas directly at virtually feed pressure. The oxygen-rich permeate stream may also find use in the plant, although this stream will probably require compression after separation. A schematic process is shown in Figure 2.

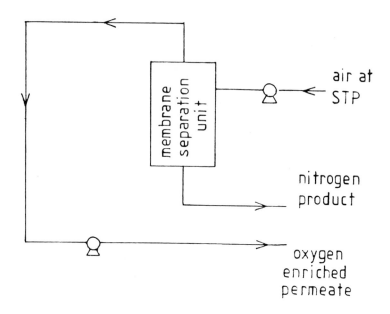

FIG.2 AIR SEPARATION

In a variant on this scheme, the efficiency of nitrogen separation can be increased by using O_2-depleted air from turbine exhaust gas or boiler flue gas as the feed to the membrane, enabling a smaller membrane installation to be used.

Nitrogen is also used in EOR applications in the same way as CO_2. In this context, membrane nitrogen separation plant could be used offshore, and an offshore nitrogen production facility based on membrane technology has been the subject of a recent design study carried out by Norsk Hydro and a subsidiary of Monsanto[11]. This study describes a proposed plant based on the Monsanto Nitrosep system, capable of producing 17×10^6 std. m^3/d of N_2 from turbine exhaust gas, to be sited on a ship permanently moored to a production platform. Naturally, membrane technology could also be used to recover methane from the associated gas produced during the stimulated oil extraction should this prove to be economic.

The separation of N_2 gives rise to an oxygen-enriched permeate stream which can also be put to use to save energy costs. Osaka Gas have developed a membrane system to provide a stream in which the oxygen content is increased to 26 - 30%[12]. This is then fed to a gas-fired furnace, resulting in improved combustion. This is especially relevant to high temperature furnaces used in the firing of ceramics and similar applications. With the present emphasis on energy conservation (and the legislative requirement for British Gas to promote conservation), the design of improved industrial furnaces may prove to be another growth area for membrane technology.

Most membranes used for oxygen enrichment have low $O_2 : N_2$ separation factors, e.g. the Osaka Gas membrane claims $\alpha_{N_2}^{O_2} \approx 2.5$. Others, such as the Envirogenics 'Gasep' membrane claim slightly higher values of about 6. Separation factors of this order indicate that large multistage separation units would be required to produce purified O_2, and this would not be competitive with cryogenic separation. However, if the separation factor could be improved, a membrane separation unit could find application in conjunction with the British Gas/Lurgi Slagging Gasifier. In this process, coal is gasified in a stream of steam and oxygen, the oxygen being provided at present by cryogenic separation. The Slagging Gasifier is a highly versatile reactor, producing a mixture of H_2, CO, CH_4 and CO_2. The cleaned and dried gas is suitable for use as industrial fuel gas, or for methanation to produce SNG, or, after treatment, as a synthesis gas for

methanol production or Fischer-Tropsch reactions. The Slagging
Gasifier may also be used as the basis of 'petrol from coal'
production, with methanol derived from slagging gasification being
converted to petroleum by, for example, the Mobil process using
ZSM5 zeolite catalysts.

One application of the Slagging Gasifier that may become
important in future is in electricity generation, as shown in
Figure 3. As stated, the product gases from the Slagging Gasifier

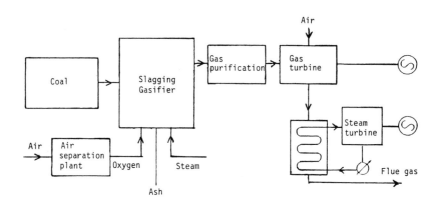

FIG.3   POWER GENERATION WITH COMBINED
CYCLE AND SLAGGING GASIFICATION

are suitable for use as industrial fuel and can be burnt in gas
turbines to produce electricity. In the combined cycle scheme
shown, the exhaust gases from the gas turbine are used to raise
steam for further power generation, thus increasing the overall

efficiency of power generation. Development of this technology would increase the potential for an efficient membrane-based O_2 separation system.

Other Applications

Apart from these three main areas, there are several other applications to which suitable membrane separation units could be put in the gas industry.

Separation of H_2S from gas streams is the most important of these. The British Gas Stretford process which produces elemental sulphur as a valuable by-product has been sold worldwide for this purpose. A membrane system which left a reject stream containing less than 2 ppm of H_2S, would enable a much smaller Stretford plant to be used, as the Stretford process would now be required to handle only the permeate stream. A combined plant such as this would probably be more economical to run than a plant using the Stretford process alone. However, gas streams encountered in the U.K. tend to have only small amounts of H_2S; without an improvement in the separating power of the commercially available membrane, this separation is only likely to be achieved using a very large membrane separation unit, and the capital cost would be prohibitive. In the chemical industry, however, many process streams are high in H_2S, and the economics of membrane separation of H_2S may be much more attractive.

Membrane technology can also be used to recover He from natural gases[13]. In Canada, Alberta Helium Ltd. have operated a multistage pilot plant to increase the He concentration of a natural gas stream to 60% from an initial 0.1%. Alberta Helium estimate that known natural gas reserves in Alberta contain

approximately 10^9 m^3 of helium. Some North Sea fields also have
low concentrations of helium and it may prove economic to recover
this.

Membranes for the separation of N_2 and CH_4 could also find
applications in treatment of well-head gases in oil fields using
N_2 in EOR schemes. Also, as some gas fields contain appreciable
amounts of N_2, removal of this N_2 using membrane technology might
also become attractive for similar reasons to those discussed for
CO_2.

Work at LRS

A research and development programme at the London Research
Station of British Gas is aimed at developing suitable membrane
systems for the applications discussed above. This entails work
in three main areas : studies of gas transport properties,
laboratory-scale testing of hollow fibre separators, and pilot
plant studies of separator performance and reliability.

Polymer films are used to measure the gas transport properties of potential membrane materials, using the Daynes-Barrer
time-lag method. Once a polymer has been selected as suitable for
further study, it is used to produce hollow fibres, and a
laboratory-scale permeator is prepared. This is then used to
study the permeability and selectivity of the membrane using both
single gases and gas mixtures. The apparatus is also used to
study the effects of various surface coatings on the properties of
the membrane.

Laboratory tests give little indication of how well a
membrane will function under plant conditions where trace
impurities might, for example, coat the fibre surface over a

period of time, reducing the selectivity, or even degrading the membrane. Accordingly, British Gas have constructed a skid-mounted testing rig capable of handling gas flows of 3 std. m^3 per hour. The pressure and flow-rate in each stream are continuously monitored, and the composition of each stream is analysed in succession by gas chromatography. This equipment allows long-term studies of membrane stability to be made and any decline in the efficiency of the separator during its lifetime to be observed.

Conclusions

In summary, the gas industry is interested in the development and deployment of membrane technology to effect hydrogen recovery in SNG production, separation of CO_2 from CH_4 in SNG and in offshore gas extraction, and also in offshore EOR applications using CO_2 miscible flood. In common with much of chemical industry, it is interested in separation of N_2 from air for use in purge gases (and also in possible offshore EOR applications) and in O_2 enrichment for improved combustion and for use in the gasification of coal.

Acknowledgements

Thanks are due to British Gas Plc for permission to publish this paper, and to our colleagues within British Gas Plc for their assistance and helpful comments in the preparation of this paper.

References
[1] T.I. Williams, "A History of the British Gas Industry", Oxford University Press, Oxford, 1981, Chapter 3.
[2] T. Graham, *Phil. Mag.*, **32**, 402, (1866).

3 S. Weller and W.A. Steiner, J. Appl. Phys., 21, 279, (1950).

4 R.W. Naylor and P.O. Backer, A.I. Chem. E.J., 1, 95, (1955).

5 C.T. Blaisdell and K. Kammermeyer, Chem. Eng. Sci., 28, 1249, (1973).

6 S.T. Hwang and K. Kammermeyer, "Membranes in Separations", Wiley, New York, 1975, Chapter 13.

7 S. Loeb and S. Sourirajan, Report No. 60 - 60, University of California, Los Angeles, 1960.

8 C.S. Goddin, Hydrocarbon Processing, May 1982, pages 125 - 130.

9 T.E. Cooley and W.L. Dethloff, Chemical Engineering Progress, 81, 45, (1985).

10 R.L. Schendel and J.D. Seymour, "Expanding Use of Membranes for CO_2 removal", paper presented at 'PetroEnergy 84', Houston, Texas, 1984.

11 T. Johanneson, 3rd European Conference of the Gas Products Association, Den Haag, 1986, paper 8.

12 Osaka Gas Technical Report No. 1 82.10, 1982.

13 C.Y. Pan and H.W. Habgood, Can. J. Chem. Eng., 56, 210, (1978).

Separation of Acid Gas and Hydrocarbons

By R. L. Schendel

FLUOR TECHNOLOGY, INC., 3333 MICHELSON DRIVE, IRVINE, CA 92730, USA

Over the last few years, membrane applications have broadened
considerably and membranes are now used in a great variety of
industries. This paper reviews the characteristics of membrane systems used for gas separations, and specifically
looks at applications for the CO_2/hydrocarbon separation as
applied in industry.

Semi-permeable membranes have been used for many years but
primarily in liquid applications such as reverse osmosis for
desalination of water and ultrafiltration for recovery of dyes
in the textile industry. More recently, semi-permeable membranes have found commercial application in the separation of
gases. Monsanto is largely responsible for the commercial
success of gas separation membranes. They used these membranes for recovery of hydrogen in their own ammonia plants
for several years before introducing the product to industry.
Other suppliers of commercial gas separation membranes include
Dow, Dupont, Grace, and Separex, with more recent announce-
ments by Union Carbide and Ube.

Membranes are thin films of any one of a number of polymers
which are specially prepared and suitable for a particular
application. Commercially available gas separation membranes
used for Acid Gas Separations have been primarily: polysulfone
and cellulose acetate.

The polymers forming the membrane may be manufactured in either
flat-sheet or hollow-fiber form. In the case of the hollow
fiber, many parallel hollow fibers are packaged together in a
manner analogous to a sheet-and-tube heat exchanger (Fig. 1).

HOLLOW FIBER MEMBRANE MODULE

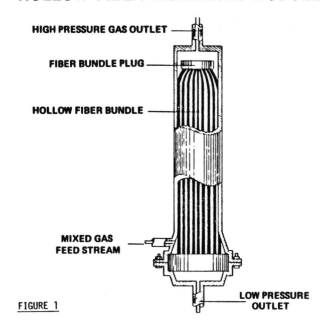

HIGH PRESSURE GAS OUTLET

FIBER BUNDLE PLUG

HOLLOW FIBER BUNDLE

MIXED GAS
FEED STREAM

LOW PRESSURE
OUTLET

FIGURE 1

SPIRAL WOUND MEMBRANE ELEMENT

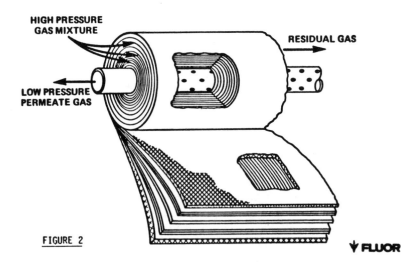

HIGH PRESSURE
GAS MIXTURE

RESIDUAL GAS

LOW PRESSURE
PERMEATE GAS

FIGURE 2

✦ FLUOR

In order to increase packing density, flat-sheet membranes are produced in spiral-wound modules (Fig.2).

The net result for either hollow-fiber or spiral-wound modules is a small package containing a large surface area of membrane. Of the four gas membrane systems available, two are hollow-fiber type and two are spiral-wound.

Characteristics

The driving force for permeation of the fast gas (and therefore separation of the fast gas from the other slow components) is the difference in partial pressure from one side of the membrane to the other.

The driving force is greatly diminished as the fast gas is removed and the partial pressure is reduced. The amount of permeate produced for the same small increment of area is much less when the partial pressure is low.

For example, with a feed gas at 500 psia and a CO_2 permeate at 50 psia, it takes almost ten times as much membrane area to allow one mole/hour of CO_2 to pass at 10% CO_2 in the feed as it does at 70% CO_2 in the feed.

On the other hand, the partial pressure and therefore driving force of the slow gas is increasing as the fast gas is removed. At low concentration of fast gas, not only is more membrane area required, but the loss of other components becomes significant.

Therefore, very pure compositions are not economically or practically attained, and membranes are not the process of choice under normal circumstances at very low concentration of fast gases.

To overcome the potential loss of desired product, membranes are used for bulk removal followed by more conventional

processes. Staging of membranes is also possible and used in
smaller systems.

Membrane systems are simple. They do not have a great deal of
associated hardware; there are no moving parts, and this is
usually an advantage.

They are modular in nature. That is, there is no significant
economy of scale, so they will tend to be more attractive when
processing lower flow rates than larger flow rates. (Most
conventional technologies do realize an economy of scale.)

A great deal of membrane area is typically packaged in a small
volume. Therefore, the entire membrane plant usually requires
less space than conventional processing.

Because membranes are simple and have no moving parts, start-
up and operation of a membrane facility is rather straightforward.

Care must be taken in the design, start-up and operation, to
protect membranes from contaminants, which could have a dele-
terious effect on the life of the membrane surface.

There are differences in the characteristics of the commercial
membranes available as well, and these may come into play in
the selection and design of the overall process. Cellulose
acetate membranes enjoy higher selectivity between CO_2 and
methane than polysulfone; therefore, a cleaner separation is
possible. In other words, the methane recovery will be higher.
However, polysulfone enjoys a distinct temperature advantage
in that the polysulfone membrane may be operated at close to
200°F. Some of the newer membranes being introduced have even
higher temperature stability.

This is particularly important when treating associated gases
with heavy hydrocarbon content.

In order to avoid condensation of heavier hydrocarbons or
natural gas liquids during CO_2 removal, it is normally necessary to pretreat the gas before the membrane separation. In
the case of cellulose acetate membranes, the gas may be chilled
to condense out the heavier hydrocarbons and then warmed back
up before feeding to the membrane unit. Another option is to
heat the gas up so that the hydrocarbon dew pont is not reached
even after CO_2 removal. The problem with this approach is the
temperature limitation of the membrane. With polysulfone
operated at the higher temperature, the second approach is
possible and advantageous. It is possible to take gas directly
from the compressor discharge and feed it to the membrane
separation. The higher temperature also allows more CO_2 gas
to permeate the membrane since the permeation rate is a function
of temperature; therefore, less membrane area will be required.

Applications

"Fast gases" will permeate the same membrane more readily than
"slow gases" with an equal driving force (Fig. 3). Hydrogen,
helium, and water vapor are considered very fast gases. That
is, they will travel through the membrane much more readily
than other gases. Moderately fast gases include the acid
gases, carbon dioxide and hydrogen sulfide. Slow gases which
tend to remain behind and not permeate the membrane include
the aliphatic hydrocarbons, nitrogen, and argon. It is, therefore, not surprising that the first applications of these gas
separation membranes have been the recovery of hydrogen, a
fast gas, from purge streams in the production of ammonia
which contain nitrogen and argon, slow gases; and in refinery
applications where hydrogen is recovered from hydrocarbon
streams. Much of the optimism for growth in the membrane
industry is based on projected hydrogen usage in future years
due to heavy and/or sour crude processing.

One area which is an attractive, but somewhat elusive, market
for membranes, and of special interest to the gas industry, is
separation of acid gases from hydrocarbon streams, specifically,
the separation of CO_2 from methane. CO_2 is only a moderately
fast gas, but the volume of gas to be treated for CO_2 removal

HIERARCHY OF PERMEATION RATES

N₂, CH₄, CO, Ar	O₂	CO₂ H₂S	He, H₂, H₂O

$N_2, CH_4, CO, Ar \qquad\qquad O_2 \qquad\qquad CO_2 \quad H_2S \qquad He, H_2, H_2O$

SLOW   **FAST**

FIGURE 3

can be huge. Some of the areas where membranes have found
commercial application for separation of CO_2 and hydrocarbons
are reviewed below.

Certain type gas and oil wells are suitable for increased production by fracturing. In the fracturing process, high-
pressure fluids are injected into the well reservoir to swell
and fracture the formation. Next, a slurry of sand is fed
into the well to fill the fracture. This forms a highly
porous channel for gas and oil to flow to production wells.
Carbon dioxide has found use as the pressurizing fluid for gas
and oil well fractures.

Fracture treatments using CO_2 are boosting production from
tight oil and gas sands in North Louisiana, South Arkansas and
East Texas. The increase in production, after fracs, has
averaged 6 times for oil wells and 3-4 times for gas wells.
The payout time for a CO_2 frac project averages one-and-a-half
months. The associated gas immediately following CO_2 fracture
necessarily contains large concentrations of carbon dioxide.
The concentration diminishes rather rapidly so, for example,
in one project, the initial CO_2 concentration one day after
the fracture was 50 to 70% CO_2 . Within a week, the concentration of CO_2 had reduced to approximately 10%, and more slowly
thereafter, until reaching levels suitable for pipeline
transmission.

Membranes are excellent for treating these associated gases
because of their modular nature and portability. Immediately
following a CO_2 frac, membranes may be used to remove CO_2 from
methane in the associated gas and as the CO_2 content comes
back down, the membranes may be removed and used elsewhere.

Separex reports at least two portable membrane systems are in use for this application, and one of these has already been used at three sites.

The CO_2 has some methane in it and is typically burned for fuel or flared.

One of the newer sources of methane gas comes from landfill and also digester gas. Both of these gases are approximately 50% carbon dioxide and 50% methane. This high CO_2 concentration and low volume of gas lends itself well to membrane processing. Monsanto reports one plant in Alabama which treats 100,000 SCFD and upgrades the gas from 600 BTU/SCF to 960 BTU/SCF.

Separex reports use of their membrane at a landfill operation processing 1½-2 MM SCFD. Feed gas concentration ranges from 42-44% carbon dioxide with as much as 2-3% air. CO_2 is removed to provide a gas with a heat content of 900-950 BTU/SCF. In this particular plant, two stages of membranes are used in series with a recycle of the permeate from the second stage to the feed compressor suction. The gas is pretreated in two beds of carbon and a coalescing filter separator before entering the first stage of the membranes. This plant has been operating since August of 1984.

Sometimes gas will be produced which is not acceptable for pipeline transmission, but if the CO_2 concentration can be reduced slightly, the gas will meet specifications.

Sun Exploration & Production Company reports membranes to be the least cost option for their operation at the Baxterville field (Mississippi, USA). They process 575,000 SCFD and reduce the CO_2 content from 5% to 2% specification.

The largest potential application for CO_2/Hydrocarbon gas separation membranes today is in the processing of gases associated with CO_2 miscible flood for enhanced oil production.

The April 1986 Oil and Gas Journal "Production/Enhanced Recovery Report" indicates approximately 30,000 BBL/day of incremental oil production by CO_2 flood in the USA.

The gases associated with this stimulation technique vary
greatly in composition and volume over the life of the project.

In many instances, existing gas processing facilities for acid
gas removal and NGL recovery are available but incapable of
handling the increased volume and high CO_2 concentration.

The carbon dioxide concentration in the associated gases can
increase to levels as high as 90 percent in as short a period
as 6 months, although carbon dioxide breakthrough within 3 to
5 years is more likely. This means the gas processor has to
contend with gases containing 80-90 percent carbon dioxide and
5 to 10 times the volume of gas in the space of 2 to 3 years!
This rise in gas volume has a profound effect on gas gathering
and treatment.

Membranes are an excellent candidate for removing carbon
dioxide from methane at the high concentration levels. Also,
due to their modular nature, membranes can be added, as
required, as the CO_2 concentration rises. The CO_2 can be
produced at intermediate pressure to reduce compression costs
for reinjection. Therefore, membranes can be effectively used
for bulk removal of CO_2 so that the remaining gas can be processed in existing equipment. In fact, this option is already
being chosen by several CO_2 flood operators.

Union Texas Petroleum has begun injection in Texas with 10 MM
SCFD of CO_2 purchased from the Sheep Mountain pipeline. They
use Monsanto membranes for CO_2 removal from the associated gas
upstream of an existing amine unit and cryogenic NGL recovery
facility.

SACROC is the only large commercial-scale CO_2 flood project
with a significant history. This CO_2 flood project has been
in operation since 1972. After pilot testing membranes for
over one year at the project, two hollow-fiber membrane plants
were installed. These two units are owned and operated by
CYNARA, a Dow subsidiary. The units handle 50 MM SCFD and

20 MM SCFD of associated gas containing 40 to 70% CO_2. Membranes are used for bulk removal of CO_2 upstream of the hot potassium carbonate units. The CO_2 product from the membranes is reinjected into the field. Start-up of these units was completed early 1984, and the units continue to operate satisfactorily.

AMOCO is using Monsanto membranes at their Central Mallet Unit in the Slaughter field. Membranes are used to remove approximately 60% of the CO_2 from produced gas which is expected to reach a volume of 100 MM SCFD and a concentration of 88% CO_2. The gas is compressed and air cooled to knock out hydrocarbons and prevent condensation in the membranes before reheating to 180-190° F. AMOCO was able to stage their capital expenditures by installing membranes eight months after initial completion.

Fluor has developed a combination process which uses both membranes and distillation to advantage. Membranes are used first to remove the bulk quantity of CO_2 and then used in a distillation scheme to remove carbon dioxide from the ethane CO_2 azeotrope (Fig. 4).

Use of membranes to remove CO_2 from the azeotrope is an excellent application of membranes since the concentration of CO_2 is quite high (approximately 65%), and the separation factor for CO_2 relative to ethane is substantially better than the separation factor for CO_2 in methane. The use of membranes as a bulk CO_2 removal device upstream of distillation also has an advantage in that it greatly reduces the turndown problems in operating a plant which must be designed for peak capacity, yet during early stages must be operated at a small fraction of the volume throughput.

Caveat

While it is true there have been two rather significant plants constructed and now in operation, using membranes, to process gas from CO_2 floods (the Chevron SACROC Plant and AMOCO Central

FIGURE 4
FLUOR – MEMBRANE DISTILLATION PROCESS

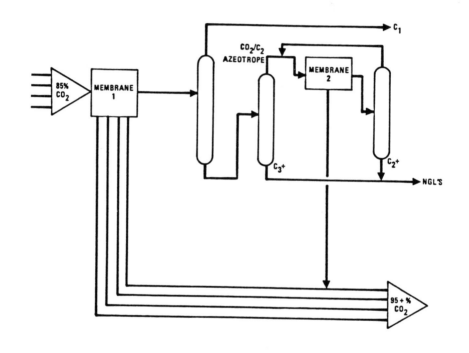

Mallet Plant) and one smaller plant (Union Texas Petroleum),
the competition for processing by distillation techniques
(i.e., Ryan-Holmes type process) is very strong. In fact,
economics lean toward the distillation process for higher
volume gases. At SACROC, high capacity hot potassium carbonate units were in place and in operation to handle
relatively high CO_2 concentrations prior to the installation
of the Dow membranes. The membranes were used essentially to
increase capacity of an already operational plant.

The AMOCO unit was also unique in the sense that AMOCO has
contractual obligations to provide hydrocarbons to an existing
NGL Recovery Plant. AMOCO, as a company, is in fact using
distillation at their CO_2 flood in the Wasson field and, by
their own analysis, expect future plants to use distillation
techniques.

A study presented in the August, 1983 issue of Hydrocarbon
Processing (special membrane issue) demonstrated use of
Fluor's Combination process to have economic advantage over
straight distillation. However, the economic advantage was
not great, and given the familiarity of operators with
distillation-type equipment, and their unfamiliarity with the
relatively new technology of gas separation membranes, this
approach has not been used commercially.

One of the major cost items is the cost of the initial
inventory of membranes themselves and their replacement. A
point to be made here is processing gases from CO_2 floods may
provide a very large and substantial market for gas separation
membranes. However, at their current cost, straight distillation techniques will continue to be preferred except in certain special circumstances.

Even when the recovery of natural gas liquids (as in the case
of associated gas from CO_2 floods) is not a factor, economies
of scale can work against the use of membrane systems for
separation of acid gases from hydrocarbons.

As reported in the Oil and Gas Journal (February 18, 1985), Fluor recently had the opportunity to do a screening study for Petroleum Corporation of New Zealand to evaluate gas pretreatment options.

Gas, containing approximately 44 mole percent CO_2 (Table 1), is received by pipeline from raw gas conditioning facilities. The CO_2 removal system is designed to reduce the CO_2 content of gas to 5.5 mole percent. By-product carbon dioxide is used for methanol plant feed.

The primary processes considered were Membranes, Membranes plus Benfield and a tertiary amine, TEA.

Membranes were thought to be an excellent candidate for several reasons:

1.   Leakage of methane into the CO_2 was not a deterrent since CO_2 and methane would be fed to the methanol plant together. The CO_2 increases the carbon content of the feed and improves efficiency of methanol production.

2.   The relatively high concentration of CO_2 (40%) in the feed gas was typical of gases which were economically treated by membranes.

3.   The client had visited the SACROC facility and was impressed by the operation of membranes there.

For more details, refer to the Oil & Gas Journal article.

Capital and operating cost estimates were prepared for the purpose of screening the cases to select one for final evaluation. Screening quality capital cost estimates were made for each case based primarily on membrane system costs from Cynara, supplier of Dow membrane systems, and Benfield and TEA system costs from Union Carbide.

TABLE 1

DESIGN BASIS

A. Gas Feed

Component	Lb Mol/Hr	Mol %
N_2	32.6	0.29
CO_2	5066.6	44.29
C_1	5007.1	43.76
C_2	711.3	6.22
C_3	410.1	3.58
iC_4	90.4	0.79
nC_4	71.2	0.62
iC_5	19.2	0.17
nC_5	12.6	0.11
C_6	16.3	0.14
C_7+	3.7	0.03
	11441.1	100.00

P = 465 psia

T = 55°F

B. CO_2 Byproduct

P = 550 psia

Min. Purity = 80 mole percent

C. Treated Gas

P = 615 psia

CO_2 Content = 5.5 mole percent

Utility requirements were treated as operating costs. Membrane life was estimated conservatively at 3 years, while membrane system maintenance costs were neglected on the basis that they would be small compared with membrane replacement costs.

All cases were then compared on the basis of total evaluated cost, defined as capital cost plus four years of operating cost. A summary of the screening estimate results is presented in Table 2.

The most dramatic difference is between the membrane only case (with recycle) and the combination of membranes and Benfield. This again points out the rapid decline in performance as concentration of the fast gas is reduced.

However, the study showed an advantage for TEA over even the best membrane case, (membranes followed by Benfield). Today, even further advantage can be realized by use of activated MDEA (another tertiary amine) processes offered by BASF and Union Carbide.

Published studies and studies by Fluor have shown the opposite results from the New Zealand situation when looking at gases associated with CO_2 flood projects. Differences to cause this result were identified.

1. EOR projects examined typically have a higher CO_2 content (50-90% CO_2). Membranes look more attractive at higher concentrations. TEA is at about maximum loading for this case.

2. Control of hydrocarbon dew point during CO_2 removal can contribute significantly to costs. The feed gas has already had sufficient liquids removed and dew point control is not part of this study.

TABLE 2

SCREENING EVALUATION SUMMARY

	Case 1		Case 2	Case 3
	80%/Membrane	90%/Membrane	Membrane/ Benfield	TEA
CAPITAL COST[1]	1.6	1.8	1.0	1.0
OPERATING COST	1.9	1.6	1.3	1.0
EVALUATED COST	1.7	1.7	1.1	1.0

Capital Cost = Battery Limits Installed Cost, Jan., 1984 U.S.G.C.

Basis, + License Fees + Initial Chemical Inventory

Operating Cost = Yearly Full Load Cost Based On:

Power @ 2.0¢ (US)/KWH

Steam @ 3.90 USD/Ton

C_3 + H.C. Loss @ 152 USD/Ton

Estimated Solvent Losses

3-Year Membrane Life

Maintenance Cost @ 3% of Installed Capital

(excluding membrane systems) per year

Evaluated Cost = Capital Cost + 4 Years Operating Cost @ Full Load

[1]All costs are shown relative to the Case 3, TEA system costs.

3. The volume of hydrocarbons being processed is much higher
    for this study than in EOR cases. Membranes do not enjoy
    economies of scale; conventional (TEA) processing does.
    This factor alone could shift the results.

A sensitivity analysis was run to examine effects of certain
assumptions used for this case study.

The 3-year life for membranes was extended for 5 years.
Capital for the Benfield unit was reduced to take
advantage of larger Benfield unit available for downstream processing and capital was added to the TEA case
for final CO_2 cleanup with Benfield.

TEA still showed an advantage!

Effect of Size

Next the effect of gas volume was examined.

EOR projects studied to date have large CO_2 concentrations and
volumes over 100 MM SCFD. However, if we look at hydrocarbon
content, the largest projects have 5-15 MM SCFD of hydrocarbon.
SACROC is larger but already committed to hot pot units in
place. The study gas had about 60 MM SCFD of hydrocarbons or
a factor of about 10 over typical EOR applications.

We extracted the costs and very roughly adjusted them for a
1/10 flow rate with the following factors:

Item	Factor
Membrane System	$(1/10)^{1.0}$
Compressor	$(1/10)^{0.85}$
Process Equipment	$(1/10)^{0.65}$
Initial Inventory	$(1/10)^{1.0}$

The best membrane case (membranes plus Benfield) now had a
slight advantage over TEA.

The results of this study are therefore in agreement with past studies by Fluor and those published in the literature.

This study also agrees with AMOCO's design philosophy at the Central Mallet unit where gas is first processed by membranes, then by a tertiary amine (activated MDEA) before finally going to a conventional amine plant for final CO_2 removal.

Conclusion

Membranes will clearly find use for carbon dioxide/hydrocarbon separations in the areas where they fit best, namely, low volume applications, temporary or short term installations, or where size and weight savings will provide a significant premium. For moderate and larger installations, they will be used in combination with other processes. To achieve a substantial sustained growth in large volume gas processing applications, the cost of processing with membranes will have to come down. This may come from an actual reduction in price or a substantial increase in performance, so that the net system cost to the operator is reduced. Only then will they compete effectively with the distillation schemes being employed today.

When membranes are being considered for a project, they should be considered in combination with other suitable processes and not only as a stand-alone process. This can lead to a large number of options. Fortunately, membrane systems can be readily simulated on the computer for rapid screening of these options.

References

[1]W.T. Jones and E. Nolley, "Surface Facilities for Processing CO_2" presented at First Annual National Enhanced Oil Recovery Conference, February 17-19, 1986

[2]R. L. Schendel, C. L. Mariz and J. Y. Mak, "Is Permeation Competitive?" Hydrocarbon Processing, August 1983

[3]J. J. Marquez, Jr. and R. J. Hamaker, "Development of Membrane Performance During SACROC Operations" presented at AIChE 1986 Spring National Meeting, April 6-10

[4]D. E. Beccue and F. S. Eisen, "Commercial Experience with a Membrane Unit for Removal of CO_2 from Natural Gas" presented at AIChE 1986 Spring National Meeting, April 6-10

[5]R. L. Schendel and J. Seymour, "Take Care in Picking Membranes to Combine with Other Processes for CO_2 Removal" Oil & Gas Journal, February 18, 1985

[6]J. Leonard, "Increased Rate of EOR Brightens Outlook" (Production/Enhanced Recovery Report) Oil & Gas Journal, April 14, 1986

[7]R. L. Schendel and E. Nooley, "Commercial Practice in Processing Gases Associated with CO_2 Floods" presented at the World Oil and Gas Show and Conference, Dallas, Texas, June 4-7, 1984

The Recovery of Helium from Diving Gas with Membranes

By K.-V. Peinemann*, K. Ohlrogge, and H.-D. Knauth

GKSS-RESEARCH CENTER GEESTHACHT GMBH, 2054 GEESTHACHT, FRG

1. Introduction

The importance of offshore technology is growing rapidly due to the development of oil- and gas fields to guarantee a constant energy supply. Therefore,
the tendency is to carry out more work at greater water depth.
   Today the main working depth is in the range of 250 meter but within the
next few years depth of 450 meter will be reached in the north sea regions.
   Underwater jobs in the offshore region involve considerable costs and
risks; they are subject to changing weather and wave conditions. By means of
a simulator it is possible to conduct tests under reproducible conditions and
relatively low cost. The GKSS underwater simulator GUSI is one of the largest
installations of this kind in the world and permits manned diving tests down
to a simulated water depth of 600 m and unmanned tests down to 2000 m of
simulated water depth. Because the pressure increases by one bar for each
10 m, the diver has to live under a significant hydrostatic pressure at a
water depth of several hundred meter. One of the many problems encountered
for men when exposed to high atmospheric pressures is the toxicity of oxygen
and nitrogen at high partial pressures. For long exposure durations the oxygen partial pressure should not exceed 0.5 bar to avoid oxygen poisoning [1].
Nitrogen exerts a narcosis when the pressure is higher than about three bar.
A standard breathing mixture for water depths of hundred meter or more is a
mixture of helium, oxygen and nitrogen (trimix). Table 1 shows typical gas
compositions for different diving depths.

	Gas Composition		
Depth (m)	% N_2	% O_2	% He
100	5	5	90
200	5	2.5	92.5
400	5	1.2	93.8

Table 1: Typical Gas Compositions for Different Diving Depths

Due to welding experiments, where argon is used as a shielding gas, the atmosphere can be contaminated with significant amounts of argon. This argon has
to be removed because it has a toxic effect at high partial pressures. Argon
represents a problem for the cryogenic plant, which is currently used for the
helium purification. It has a relatively high melting point (- 189.2 °C) and

is solid at the temperature of liquid nitrogen (- 195.8 °C). Therefore, the icy argon may block the cryogenic plant.
The purity requirements of the reclaimed helium are shown in table 2.

Helium	Purity	Requirements
He		99.8 Vol. %
O_2	max.	0.5 Vol. %
N_2	max.	0.5 Vol. %
CO_2	max.	10 ppmv
CO	max.	2 ppmv
Ar	max.	0.2 Vol. %

Table 2: Purity requirements of the reclaimed helium

The goal of our current helium purification project is to develop a membrane based purification system, which may represent an alternative to the cryogenetic helium recovery.

GUSI requires 6000 m³ of helium for one dive to a water depth of 300 m just in order to fill the pressure chambers. In an offshore dive of 30 days at a depth of 300 m involving six divers working two hours each daily the total consumption would be 38000 m³ of helium (representing a value of about 600.000 DM) if no helium recovery and purification systems were available. In order to reduce these operating costs helium recovery and purification systems for the chamber gas and the breathing gas of the divers are used in underwater operations. The composition of the contaminated gas depends on diving depth and working conditions and can be seen in table 3.

Gas	Concentration	
He		80-95 Vol %
O_2	max.	20 Vol %
N_2	max.	20 Vol %
CO_2	max.	1,5 Vol %
CO	max.	50 ppmV
Ar	max.	10 Vol %

Table 3: Composition of contaminated gas

2. Membrane development

Preliminary calculations showed that a membrane selectivity helium to nitrogen of at least 50 preferably 100 is necessary, if the helium loss is to be kept below 5 %. To restrict the plant dimensions to a resonable size, it was decided that a helium flux of at least 0.1 preferably 0.2 m^3/m^2 h bar is required. Table 4 shows a selection of polymers with high permeability ratios for helium to nitrogen. The selection is somewhat arbitrary; however, there are not many polymers showing He/N_2 selectivities of 100 or more. Due to their low gas permeabilities most of the high selective polymers are not useful as membrane materials. As a result of our membrane polymer screening, the synthetic resin polyetherimide (PEI) has been selected. Polyetherimide is an engineering polymer manufactured by the General Electric Co. under the trade name Ultem. It is an amorphous thermoplastic resin offering good mechanical strength and high heat resistance. The He/N_2 permeability ratio of PEI given in table 4 was determined using asymmetric silicone coated membranes. Because it is difficult to assess the effective film thickness, the permeabilities for PEI are uncertain. The given data have been calculated from the gas fluxes through asymmetric membranes and an effective film thickness of 0.2 µm, which was ascertained by electron microscopy. The asymmetric membranes have been prepared by the Loeb-Sourirajan technique.

Polymer	Permeability (25 °C) $cm^3(STP)cm/cm^2sec$ cm Hg x 10^{10}		Ratio
	P(He)	P(N_2)	He/N_2
Polyamid[+] (Nylon 6)	2.430	0.0246	98.8
Polyethylen-[+] terephtalat (Mylar)	1.002	0.006	167
amorphous film[+]	2.967	0.0144	206
Polyvinylfluoride[+]	0.970	0.0042	231
Polyvinylidenchloride[+]	0.066	0.00018	366
Polyetherimide	(20)	(0.08)	234

[+] Data from (2)

Table 4: Selected Polymers with High Permeability Ratios for He and N_2

The optimal casting solution is given in table 5.

Casting Solution for Dense Asymmetric Polyetherimide Membranes		
Polyetherimide	15.9	weight %
Dichloromethane	54.6	"
1,1,2 – Trichloroethane	4.8	"
Xylene	17.6	"
Acetic Acid	7.1	"

Table 5

Dichloromethane and trichloroethane are solvents for PEI whereas acetic acid and xylene are swelling agents. Acetone has been determined as the best precipitation liquid. With several other organic solvents (e.g. toluene) as precipitation medium very selective membranes could be obtained also but the helium fluxes were considerably lower [3]. Table 6 shows fluxes of helium, hydrogen and nitrogen through asymmetric PEI membranes.

	Flux cm³(STP)/cm²sec cm Hg x 10⁴			Flux-ratio	
	$Q(H_2)$	$Q(He)$	$Q(N_2)$	$Q(H_2/N_2)$	$Q(He/N_2)$
Uncoated membrane (20 °C, 20 bar)	0.73	1.03	0.069	5-50	5-70
Silicone coated membrane (20 °C, 20 bar)	0.70	0.90	0.005	140	180
(80 °C, 20 bar)	–	2.79	0.019	–	145

Table 6: Helium-, Hydrogen- and Nitrogen-Fluxes Through Asymmetric PEI-Membranes

To gain information on the integrity of the separating layer, silicone coating was applied. The fluxes were determined at an operating pressure of 20 bar and at temperatures of 20 and 80 °C. As can be seen from the table, coating with silicone film results in a considerable increase in He/N_2 selectivity. Silicone coating affects the helium flux only slightly, but strongly reduces the N_2 permeability. The data presented are average values obtained with laboratory samples (membrane area = 34 cm²). In the range between 5 and 100 bar the helium and nitrogen fluxes were nearly independent of presure as can be seen in Figure 1.

The temperature dependence of the gas fluxes is shown in Figures 2 and 3. Whereas the helium flux increases about threefold, when the temperature is raised from 20 bo 80 °C, the nitrogen flux increases by a factor of 5.6. This leads to a decrease in selectivity with temperature, which is shown in Fig. 4.

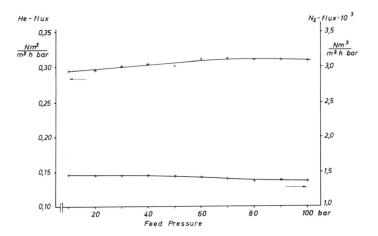

Fig. 1: He and N_2 - Flux vs Pressure

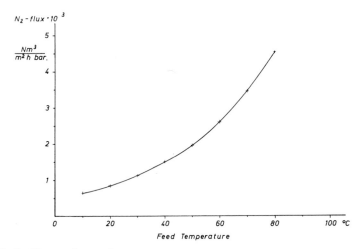

Fig. 2: N_2-Flux vs Temperature

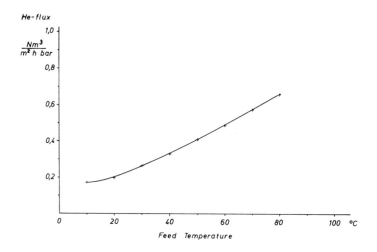

Fig. 3: He - Flux vs Temperature

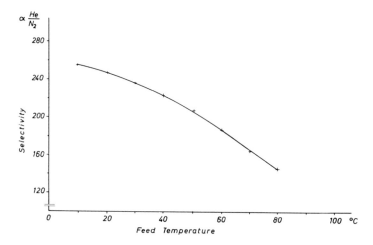

Fig. 4: He/N₂ - Selectivity vs Temperature

All data presented so far have been achieved using pure gases. To answer the question, whether permeabilities measured for pure components can be used directly in mixed-gas calculations, four different He/N_2 mixtures have been prepared; the permeation experiments were carried out at very low stage cuts. The operating pressure was 20 bar, the temperature 20 °C. The observed selectivities vs. N_2 content of the feed gas are shown in Figure 5. The helium flux dropped from .283 m^3/m^2 h bar for the 90 % He mixture to .200 m^3/m^2 h bar for the 15 % He mixture. The nitrogen flux increased from .0011 m^3/m^2 h bar for the 90 % He mixture to .002 m^3/m^2 h bar for the 15 % mixture. This is in agreement with earlier observations, that the presence of a slow gas can reduce the permeability of a fast gas, and that the presence of a fast gas can increase the permeability of a slow gas [4,5,6].

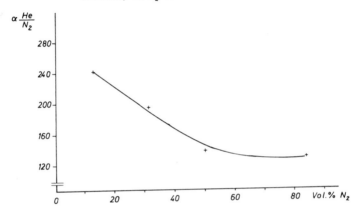

Selectivity vs. N_2 Content of Feed Gas

Fig. 5

3. Module description

Not only the membrane but also the module for the helium recovery plant has been developed at GKSS research center. A drawing of the module is shown in Fig. 6. The main parts are the pressure vessel and a number of membrane envelopes. These envelopes are consisting of two PEI membranes, which are sealed at the edges with the aid of a conventional foil sealing machine. In the inside of each envelope is a permeate spacer. One envelope has an area of 400 cm^2. The membranes are stringed on a central permeate tube and are pressed together with two endplates and two clamping sleeves. The permeate tube is open at both ends. Our standard module which is designed for a pressure of 64 bar and which will be installed in the pilot plant has an outer diameter of 21,9 cm and a total length of 28,4 cm. The effective membrane area is 4 m^2.

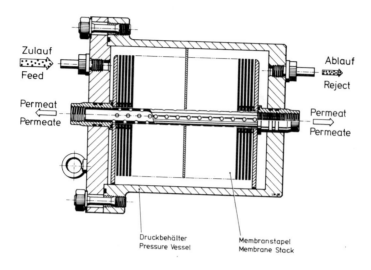

Fig. 6: GS - 2 Module

4. Membrane module evaluation

To evaluate the performance of the membrane modules under realistic conditions, the construction of a closed test loop was necessary. A flow diagram of this test loop is shown in Fig. 7.

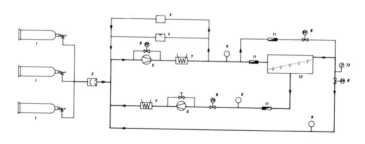

1 Gas supply   4 Gas moistening chamber   7 Heat exchanger   10 Separation module
2 Gas mixing device   5 Feed compressor   8 Pressure regulator   11 Mass flow meter
3 Gas drying chamber 6 Permeate compressor   9 Measuring point for   12 Pressure gauge
                                                   pressure, temperature,
                                                   moisture and compo-
                                                   sition of gas

Fig. 7: Gas loop

Pure gases were mixed and then fed into the gas loop. The gas was pressurized and circulated by the feed compressor (Hofer MKZ 550-10). At an outlet pressure of 64 bar the maximum flow rate of this compressor 500 Nm³/h. The permeate was recompressed by an additional compressor (Hofer MKZ 350-10) and mixed again with the retentate. Compositions of feed, retentate and permeate were analyzed with an on-line gas chromatograph (Varian Vista 6000). With this loop there is the ability to vary the following test parameters
- composition of the feed gas (2)
- feed gas pressure   (8)
- feed gas flow   (8)
- feed gas temperature   (7)
- moisture content: gas moistening (4), gas drying (3).
For example, using this test system one module with PEI membranes has been tested for 1500 hours. The helium content of the feed stream was changed between 80 and 95 % helium. The feeed pressure was 40 bar. No significant changes in gas permeabilities and selectivities have been observed.

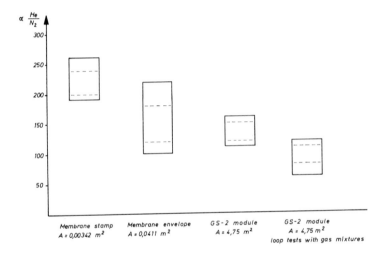

Fig. 8: Selectivity vs membrane configuration

It is decided that only modules showing He/N₂ selectivities greater than 80 are selected for the helium reclaim pilot plant. The range of module selectivities for gas mixtures containing 90 % He, 5 % O_2 and 5 % N_2 was 60 to 120 as shown in Fig. 8. The second point emerging from Fig. 8 is that the selectivity decreased for larger membrane configurations. For single membrane stamps (area 34.2 cm²) an average He/N₂ selectivity of 220 could be obtained. The average selectivity of membrane envelopes (area 411 cm²) was about 140. This decrease in selectivity can probably be attributed to very small leaks in the welding seam.

5. Process design

Using the data obtained with the module test loop a helium recovery plant for GUSI has been designed. The design parameters have been fixed as follows:

product flow	100 Nm³/h
product purity	99.8 %
feed gas composition	90 % He, 2 % O_2, 8 % N_2
membrane	polyetherimide
helium permeability	0.183 Nm³/m² h bar
average O_2/N_2 permeability	.004 Nm³/m² h bar
flow pattern	cross flow
feed pressure	60 bar

It turned out that a one stage process with recycle as shown in Fig. 9 is necessary to obtain the required helium purity and to restrict the helium loss to 0.5 % [7]. To achieve a purified helium flux of about 100 Nm³/h the membrane area in stage 1A (enriching stage) has to be 10 m². For stage 1B (stripping stage) an area of 16 m² is required.

If the helium product purity has to be increased by only 0.15 % (from 99.75 to 99.9 %) the membrane area of stage 1B has to be more than doubled, the recycle flow has to be increased 3 times as can be seen from Fig. 10. Likewise the compressor capacity has to be extended 3 times.

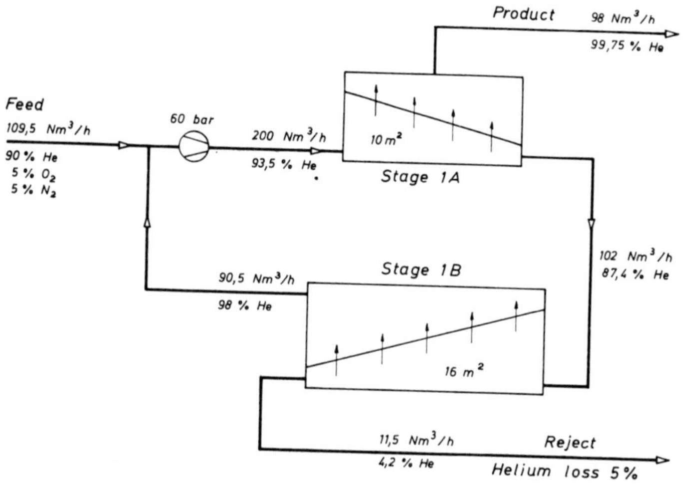

Fig. 9: Standard lay out

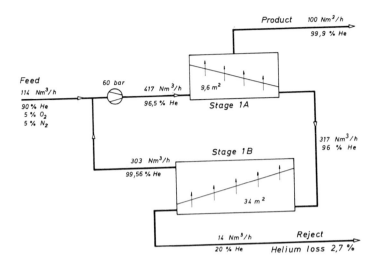

Fig. 10: Plant design for a helium purity of 99,9 %

6. Pilot Plant Description

The design of the pilot plant for helium reclaim from diving gas was fitted to the requirements of the GKSS underwater simulator.

The contaminated helium is stored in storage tanks with a pressure of 200 bar. This pressure of 200 bar was reduced to the level of 60 bar and was used as pressure supply for the gas separation system. The feedgas flow rate was controlled by a thermal mass flow controller. The lay out of the separation system is based on the principle of the process engeneering for single stage with recycle. The gas stream entered the enriching stage, the residue of this stage was separated in the stripping stage in an enriched helium stream at the permeate side and an enriched O_2/N_2 stream at the residue side. The helium concentrated permeate was recompressed by an permeate compressor to 60 bars and recycled to the feed stream to increase the helium concentration. Based on this process design the required helium purity in the permeate of the enriching stage can be achieved and the helium loss can be minimised. The system design was developed on the basis of the loop experiments performed with the GS2-module running different gas mixtures and feed gas pressures. The membrane system is a combination of 8 GS2-modules, each packed with 4 m^2 membrane area. The modules for the enriching stage and stripping stage are racked together in parallel. Depending on the feedgas concentration the number of modules can be connected to the enriching or stripping stage. The membrane area of each stage can be vary from 8 to 24 m^2.

The residue rate of the stripping stage is controlled by a thermal mass flow controller. For each separation column the data of volume flow, pressure, temperature and gas concentration of feed, residue and permeate can be measured and plotted.

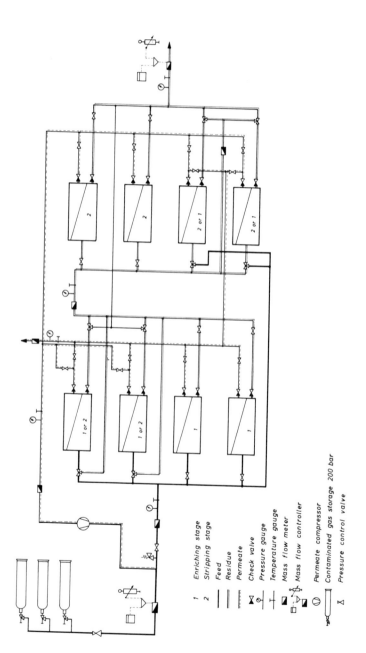

Fig. 11: Pilot Plant for Helium Reclaim for Diving Gas

1   Enriching stage
2   Stripping stage
        Feed
        Residue
        Permeate
        Check valve
        Pressure gauge
        Temperature gauge
        Mass flow meter
        Mass flow controller
        Permeate compressor
        Contaminated gas storage 200 bar
        Pressure control valve

The pilot plant is now under construction, the piping, manifolds and measuring systems are erected and the membrane stack is ready to be filled with the modules. As mentioned earlier, before a module is connected to the plant each module is tested in the gas loop for 200 hours under changing pressure and gas conditiones. The start up ot the pilot plant is planned for late 1986.

7. Reference

[1] P.B. Bennett and D.H. Elliot, "The Physiology and Medicine of Diving", Baillière Tindall, London, 1982, Chapter 9

[2] S.T. Hwang, C.K. Choi, K. Kammermeyer, *Sep. Sci.*, 1974, *9*, 461

[3] K.-V. Peinemann, DE 3420373A1 (5.12.85)

[4] D.G. Pye, H.H. Hoehn, M. Panar, *J. Appl. Polym. Sci.*, 1976, *20*, 1921

[5] F.P. McCandless, *Ind. Eng. Chem. Proc. Des. Dev.*, 1972, *11*, 470

[6] C.R. Antonson, R.U. Gardner, C.F. King, D.Y. Ko, *Ind. Eng. Chem. Proc. Des. Dev.*, 1977, *16*, 463

[7] K. Hattenbach, "Ein neues Arbeitsdiagramm für Membrantrennverfahren", GKSS 85/E/20

Generon* Air Separation Systems - Membranes in Gas Separation and Enrichment

By K. B. McReynolds

DOW CHEMICAL COMPANY, MIDLAND, MI, USA

INTRODUCTION

The use of membrane technology for air separation is now commercially available in GENERON air separation systems from The
Dow Chemical Company. The GENERON air separation system
separates ordinary compressed air into an enriched nitrogen product stream controllable up to 99% nitrogen and an enriched
oxygen vent stream controllable up to 35% oxygen. This separation is accomplished utilizing hollow-fiber membranes housed in
modules with no moving parts and no complex pretreatment.

Although the current GENERON air separation system has been
designed and optimized for production of the enriched nitrogen
product stream, both streams are recoverable simultaneously. The
nitrogen stream produced by this GENERON air separation system is
clean and dry.

Under the design conditions of 95% nitrogen/298°K (25°C) and 725
kPa (90 psig) feed pressure, GENERON air separation systems are
available in sizes up to 400 nM^3/hour (250 SCFM) of enriched
nitrogen product. The ten module system shown here will deliver
80 nM^3/hour (3000 SCFH). Systems are available up to 863 kPa
(110 psig) feed pressure.

This paper is concerned with the total operating costs of a
GENERON air separation system. The components of the costs
discussed are included in Table I. Before discussing the costs
of GENERON air separation systems, a brief description of
membrane technology is included.

*Trademark of The Dow Chemical Company

DESCRIPTION OF MODULE

As mentioned earlier, the GENERON air separation system module is
a hollow fiber membrane device designed similarly to a shell-and-
tube heat exchanger. The feed air is introduced into the module
through the centrally located distribution core. The distribution core traverses the full length of the module and it is
embedded and plugged in the non-feed tubesheet. Between the two
tubesheets are the hollow fibers, lying parallel to the distribution core.

The fibers are embedded in the tubesheets and are opened on the
endplate sides of the tubesheet (referred to as the permeate or
low pressure side of the module). The low and high pressure
sides of the module are separated by O-rings on the tubesheets.

The feed air leaves the distribution core through holes placed in
the distribution core between the two tubesheets into the area of
hollow fibers. The feed air passes radially to the pressure
case. As the air passes the hollow fibers, the oxygen is being
depleted through the hollow fiber wall at a faster rate than
nitrogen leaving an oxygen depleted nitrogen stream. The nitrogen stream is tapped off from the middle of the pressure case.
The oxygen rich permeate stream passes down the bore of the fiber
and is removed from both ends of the module.

The pressure drop across the module (distribution core to
enriched nitrogen tap) is 138 kPa (5 psig) under design conditions. Hence the nitrogen product stream exits the GENERON air
separation system at approximately 690 kPa (85 psig) under design
conditions. The oxygen rich stream exits at less than 105 kPa
(0.5 psig). The modules are only one component of the GENERON
air separation system.

TABLE I

GENERON AIR SEPARATION SYSTEM

Components

- COMPRESSOR/CHILLER (optional equipment)
- PREFILTER
- FEED AIR VALVE
- PRESSURE CONTROL VALVE
- MEMBRANES
- FLOW CONTROL VALVE
- PRODUCT DISTRIBUTION/DIVERTER VALVES
- VENT (oxygen enriched air) SYSTEM
- INSTRUMENTATION/CONTROL PANEL

SYSTEM OPERATION

To start-up and operate the skid, switch to the start position on
the instrument panel. This brings power to the system. Pushing
the reset button will open the feed air valve. During start-up,
check the feed pressure, filter drains and $\triangle$P gauge across the
prefilter. When the oxygen level in the product gas drops below
7.5% oxygen (or whatever set point maximum for oxyen, adjustable
in oxygen analyzer) the vent valve can be closed and the distribution valve can be opened. Stopping of the system is accomplished by switching from start to the stop position. This
closes the feed air valve. Leaving the power switch in the on
position allows for faster system start-up since there is a warm-
up time with the oxygen analyzer.

MODULE PERFORMANCE

The module performance is impacted mainly by the feed air
pressure and the desired nitrogen product purity. This is
demonstrated graphically below (Graph 1). At design conditions
the module performance is 8 nM^3/hr of 95% nitrogen. Increasing
feed pressure does improve module output. Closing the flow
control valve increases the feed air dwell time in the module;
this reduces the oxygen content in the product gas.

Purity vs. Product Flow
(Performance of a single GENERON module at 25°C)

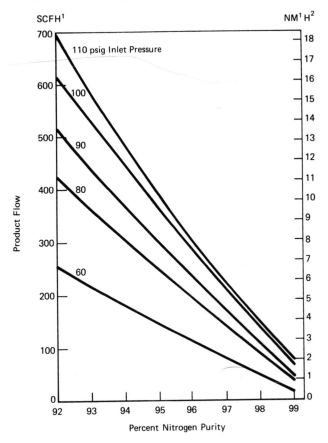

[1]Conditions of SCFH are one atmosphere of pressure at 70°F.

[2]Conditions of NM³ are one atmosphere of pressure at 0°C.

MAINTENANCE COSTS

1. Maintenance of GENERON Skid

    The GENERON air separation system for enriched nitrogen pro-
    duction has been designed for simple and inexpensive main-
    tenance. The filter element in the coalescing filter
    assembly will need periodic replacement. The coalescing
    filter assembly has been placed on the skid for easy access.
    Only minutes are needed to separate the automatic drain and
    filter assembly, unbolt and remove the pressure case and
    insert a new filter element. The filter has been over-sized
    needing only yearly replacement on typical compressed air.
    Compressor manufacturers specify typical oil flooded com-
    pressed air contains less than 3 ppm of oil. Unusually con-
    taminated feed air will necessitate more frequent element
    replacement and/or additional filtration equipment. Com-
    pressor oil vapor is a fast permeator and preferentially
    exits the module in the oxygen enriched air stream.

    The GENERON hollow fiber membranes are very tolerant to
    impurities in the feed air; however, excessive exposure to
    liquids (compressor oils and liquid water) will cause feed
    air channelling in the module. Channelling will necessitate
    restriction of product flow to maintain product purity. The
    maintenance cost on a GENERON air separation skid is typi-
    cally less than $0.001/nM^3.

2. Maintenance of Compressors

    The GENERON air separation system has been designed to
    operate using feed air produced by a single stage oil-flooded
    screw air compressor. This type of compressor is known for
    its low capital cost, easy installation, minimal operator
    attention and inexpensive maintenance. The maintenance cost
    per year is typically 5% of installed capital cost or less
    than $0.001/nM^3 of product gas.

3. Maintenance of Chiller

The chillers used in the GENERON air separation systems are compressed air driers, which have not had an economizer heat exchanger placed on the incoming feed air (typical feed air 20°C). This type of equipment normally receives minimal annual maintenance and has a use life of up to 15 years.

The chiller is used on the GENERON system for three reasons. First, sizing of the feed air compressor is impacted by the feed temperature to the modules. Maintaining a constant, cool feed temperature to the modules permits the economic sizing of the feed air compressor possible. Secondly, the chiller removes additional water liquid/vapor and compressor oil liquid/vapor reducing the risk of module exposure to liquids contamination. This helps optimize module life. Third, chilled feed air contains a constant water vapor content allowing for predicted nitrogen product pressure dew points.

POWER COSTS

1. Air Compressor

The major cost in operating the GENERON air separation skid is the power to run the air compressor. The feed flow/module is 32 nM^3/hr at design conditions. The power cost[1] to operate the compressor is $0.03/$nM^3$ of product gas produced.

2. Chiller

The power cost to chill air is extremely low. Typically less than 0.25 horsepower/module is needed to lower the feed air to 280°K (7°C). Approximately 40% less feed air is needed to operate at 280°K (7°C) than at a feed air temperature of 310°K (37°C). The power cost for the chiller is less than $0.001/$nM^3$ of product gas produced. The modules can be operated down to the freezing point of water with no problems. Operation at less than the freezing point of water is possible; however, the GENERON membrane air separation representative should be contacted for proper operation.

3. GENERON Skid

Power is needed to operate the GENERON skid itself. A 110 V,
15A power source is needed to operate control panel, the feed
solenoid valves and the oven in the zirconium oxide oxygen
analyzer. The power cost to operate the GENERON skid is
$0.001/nM^3.

CAPITAL COST

1. GENERON Skid

The GENERON air separation skid price lists (including module
replacements) are available from the local GENERON represen-
tative. The capital cost, if straight line depreciated over
10 years, is approximately $0.023/nM^3 of produced gas. This
capital cost includes module replacement over 10 years. The
GENERON systems have been designed to maximize module life.
Proper maintenance and operation of the GENERON system will
insure a minimum replacement module cost.

Proper maintenance and operational procedures have been
described in the easy to read GENERON air separation system
Operating Manual.

APPLICATION

The initial market opportunities for a GENERON generated inert
gas were blanketing, sparging and purging in various applications
in the chemical processing industry. Our first GENERON air
separation system was installed in January, 1980 in our own
acrylonitrile butadiene styrene (ABS) plant in California
blanketing styrene monomer. This unit, under design conditions,
is a 130 nM^3/hr of 95% nitrogen. This particular system is still
in operation today.

[1]Power cost at 5.5¢/Kwh

The first system installed outside of The Dow Chemical Company
was in May, 1984. This system produces 40 nM^3/hr of 95% nitrogen
using existing plant compressed air. The system was installed at
a large paint manufacturer. The enriched nitrogen product stream
is being used for various blanketing and sparging applications in
their alkyd resin manufacture. A case history will be available
later this year.

Since 1980, many other applications for GENERON generated gas
have been discovered. Uses in the food, pharmaceutical, metal
treating and paint industries are only a few of the many applications where GENERON inert gas has been accepted.

Since our commercial launch in January, 1985 over 50 systems of
various sizes have been ordered for GENERON varying in size up to
200 nM^3/hr.

Metal treating was an application area which initially was viewed
as offering little potential for GENERON atmosphere. However,
initial trials at several metal treating installations have proven that GENERON atmosphere is a viable nitrogen source in
nitrogen/methanol carburizing atmospheres.

SUMMARY

The total operating cost of a GENERON air separation system is
shown in Table II. The total cost of operation is $0.06/$nM^3$ of
gas produced (95% nitrogen). Approximately 50% of the cost is
power to operate the feed air compressor.

Operation at higher nitrogen purities are possible. At 99%
nitrogen the cost increases to $0.15/$nM^3$. At 99% nitrogen the
costs are still split evenly between power and
capital/maintenance cost.

TABLE II

TOTAL OPERATING COST/M^3 OF PRODUCT GAS

Maintenance	$0.002
Power Cost	0.032
Capital Depreciation	0.026
TOTAL	$0.06

GENERON has received an excellent initial market acceptance and
the use of membranes as a process for generation of inert gas
appears to have a very bright future.

Polyacetylene Derivatives as Membranes for the Separation of Air

By T. Nakagawa

DEPARTMENT OF INDUSTRIAL CHEMISTRY, MEIJI UNIVERSITY, TAMA-KU, KAWASAKI, JAPAN

Introduction

In recent years, considerable work has been done on gas separation by membranes [1]. Membranes used for gas separaton are required to have the following properties: a) high gas permeability coefficient, b) high separation factor which is defined as the ratio of permeability coefficients, c) formation of nonporous thin layer, and d) durability. Oxygen/nitrogen separation by membranes, which are able to achieve an oxygen enrichment of 40% concentration from air have already been commercialized for the medical use. More attractive use of membranes of oxygen enrichment for combustion, however, has proved difficult state, because of rather low oxygen permeation rate. While the silicone rubber, namely poly(dimethyl siloxane), has the highest oxygen permeability of all nonporous membranes synthesized on the industrial scale, there are a few defects, that is, the lowest separation factor and the difficulty of forming a thin membrane of which thickness is less than one micron. Therefore, in order to overcome these defects, block copolymers composed of poly(dimethyl siloxane) and polycrbonate [2], or poly(dimethyl siloxane) and

351

poly(hydroxystyrene) [3], have been used for oxygen enrichment. Compared with poly(dimethyl siloxane), the permeabilities of these copolymer membranes to oxygene decrease about one third, but the permeation rates are extremely high, because of 0.1-0.03 μm of thickness of the membranes, and the separation factors are almost the same values. The permeability coefficient of poly(dimethyl siloxane) to oxygen has been reported to be 6.0×10^{-8} $cm^3(stp) \cdot cm/cm^2 \cdot sec \cdot cmHg$ at $30°C$ [4], which differs by depending on the degree of crosslinking. This value is the highest and has been considered to be the limiting one. However, very lately a new membrane was synthesized by Masuda and Higashimura [5-7], and it was found that the permeability to oxygen is 10-50 times higher than that of poly(dimethyl siloxane) [8].

However, the author and his coworkers found that membranes of this polymer were extremely unstable in permeation behaviour and the permeability coefficients decreased with time by thermal histeresis or absorption of volatile materials in the early stage of research [9]. The objective of the present study was to clarify the un-usual behaviour of permeation and to modify the membrane to a more stable one with better separation properties.

Experimental

Polymerization of 1-(timethylsilyl)-1- propyne.

1-(trimethylsilyl)-1-propyne supplied from Shinetsu Chemical Industry Ltd. was distilled twice from calcium hydride under atomospheric pressure. The boiling point was 99-100 °C. Polymerization of 1-(trimethylsilyl)-1-propyne was carried out according to Masuda's method[7] in some parts. An example of procedure was as follows:

30 ml of anhydrous toluene was placed in a one-liter, three necked flask equipped with a mechanical stirrer, thermometer, inlet and outlet tubes for nitrogen gas and a dropping funnel, followed by bubbling with dry nitrogen gas for about half an hour. Then 0.49 g of TaCl₅ was dissolved. Finally 1-trimethylsilyl-1-propyne (monomer) was added. Temperature of the reaction mixture was raised until 60°C, and kept at 60°C for about three hours. The reaction mixture was then poured into one liter of methanol. A fibril-like polymer was precipitated. The polymer was separated by filtration and dried. The polymer was reprecipitated from a toluene solution and washed and dried under vacuum prior to determining the degree of polymerization. The conversion was about 40 %. The degree of polymerization was 1.20 X 10^6 by GPC.Poly[1-(trimethylsilyl)-1-propyne] is abbreviated as PMSP hereafter.

$$CH_3-C\equiv C-Si(CH_3)_3 \xrightarrow[\substack{Cat:TaCl_5 \\ solvent:toluene}]{\sim 60°C} -(\underset{Si(CH_3)_3}{\overset{CH_3}{C=C}})_n-$$

$$\overline{Mn} \doteqdot 1.2 \times 10^6$$

Preparation of PMSP membranes

The membranes were cast at room temperature (20-25°C) from a toluene solution on glass plate floating on mercury. The toluene was evaporated very slowly for about one week. The membranes were then immersed in methanol for several days and dried under vacuum. The resulting membrane thickness was between 30-200 μm, depending on the concentration of casting solutions.

Modification of the PMSP Membranes by additives

Modification of the membrane was carried out by using adsorption apparatus as shown in Fig. 1. In a bell-type vessel, the membrane between filter paper was put on a share in which an additive such as a plasticizer was installed. Then the vessel was evacuated and kept at vacuum and the share was heated from outside using a hot plate. The whole vessel was kept at $100°C$. The vapour evolved from the share was absorbed into the membrane, and also the membrane was simultaneously thermally treated. The weight % of additives absorbed into the membrane was calculated as follows:

$$\text{wt\% of the additive} = \frac{W - W_o}{W_o} \times 100$$

where, W_o, W are weight of the original membrane and weights of the membrane containing additives respectively. The ammounts of additives absorbed with the membrane were controlled by exposure intervals, temperature and degree of evacuation.

Permeability Measurements.

The general theory of gas transport in polymers and detailed discussions of the methods of measurement and calculation of the permeability,diffusion and solubility coefficients have been published elsewhere[10]. The experimental method used in this study was an adaptation of the high vacuum gas transmission technique wuth an MKS Baratron Model 310BHS-100S pressure transducer instead of Model 90 [11]. In the early stage of the experiment, a butyl rubber was used as a rubber gasket. Then Silastic silicon rubber with no additives was used unless otherwise mentioned.

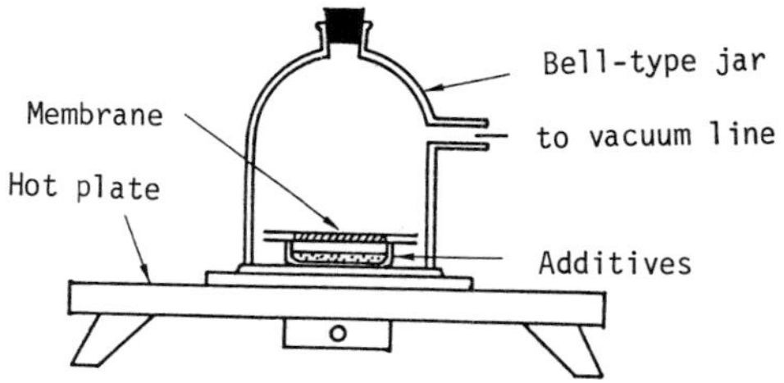

Fig.1. Adsorption Apparatus of additives to membrane.

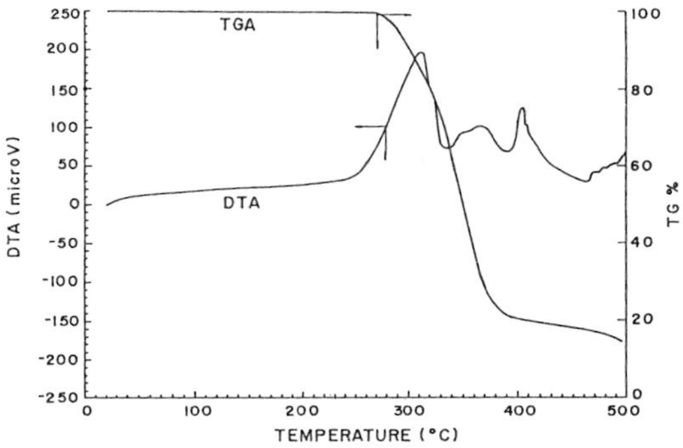

Fig.2. DTA and TGA curves of PMSP membrane.

Table 1. Permeability coefficient($\bar{P}$), effective diffusion coefficient($\bar{D}$) and permeative solubility coefficient($\bar{S}$) of PMSP membrane at 30 C.

Gas	$\bar{P} \times 10^8$	$\bar{D} \times 10^5$	$\bar{S} \times 10^3$
He	50.8	68.9	1.91
H_2	132	10.2	3.23
Ne	32.9	3.58	21.6
O_2	77.3	2.65	26.3
Ar	69.7	2.27	27.3
CH_4	130	2.55	57.3
N_2	49.7	2.17	129
CO_2	280	1.33	68.3
Kr	90.8	0.443	264
Xe	117		

$\bar{P}$: $cm^3(STP) \cdot cm/cm^2 \cdot s \cdot cmHg$
$\bar{D}$: cm^2/sec, $\bar{S}$: $cm^3(STP)/cm^3(polymer) \cdot cmHg$

X-Ray and Other Characterizations

The methods used for X-ray diffraction characterization and DTA,
TGA characterization were carried out as usual.

Results and Discussion

According to Masuda's reports [5], the results of dynamic visco-
elastic measurement showed no glass transition between -150 and
+200°C. Our data of DTA and TGA, shown in Fig. 2, also showed no
transtion between 0-280 ° C. The glass transition temperature is
considered to be higher than 200°C. Therefore, the permeability
experiment was done in a glassy state of PMSP. Takada et al. have
already reported data of gas permeability of poly(acetylene)
derivatives including PMSP. They showed permeability of PMSM to gases
was extrememly high compared with that of silicone rubber, and
negative apparent activation energy for permeation. However, no reason
was shown clearly for this and other unusual behaviour of gas
permeation. First of all, the permeability coefficients, $\bar{P}$, effective
diffusion coefficient, $\bar{D}$, calculated with time lags and permeative
solubility coefficients, $\bar{S}$, at 30°C, of PMSP membranes to gases,
which were prepared at 25-30°C and were not thermally treated, are
summarized in Table 1.

Diffusion coeficients of gases in PMSP membrane were also compared
with those in silicone rubber in Fig. 3 . Interesting to describe,
diffusion coefficients in both a rubbery state silicon rubber and a
glassy state PMSP membrane are quite similar. From Table 1 and
Fig. 3 , the higher permeability of PMSP could be correlated with the
permeative solubility coefficient. The higher diffusivity may be due to

the formation of large space between polymer segments by the bulky side chains of trimethylsilyl group. The higher solubility could be ascribed to a high content of C=C double bonds, because Bixler, et.al. have been made clear the increased solubility coefficient of γ-irradiated polyethylene by the formation of C=C double bonds [12], and the author also reported the same phenomenon by γ-irradiation followed by thermal treatment of poly(vinyl chloride) [13].

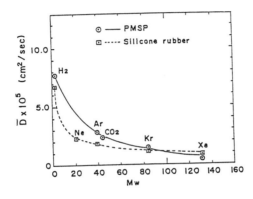

Fig.3. $\bar{D}$ in PMSP and poly(dimethyl siloxane) to gases vs. their molecular weights.

The effect of temperature on permeability or effective diffusion coefficients of N_2 and O_2 through PMSP membrane, which was thermally treated at 50°C for few days just before the permeation experiment, is shown in Fig. 4. As can be seen in Fig.4, the apparent activation energies for permeation, Ep , are negative, while those of diffusion are positive. This means that the transport of gases through PMSP membrane occurs by an activated diffusion process. According to the solution-diffusion theory, Ep is the summation of Ed and ΔHs. The

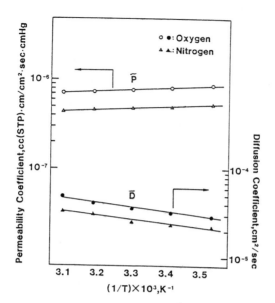

Fig.4. Temperature dependence of $\bar{P}$ and $\bar{D}$ for oxygen
and nitrogen in PMSP membrane.

negative Ep value came from the large negative value of ΔHs. Negative
ΔHs values were frequently found in polymeric membrane-gas or vapour
systems, which have high solubility coefficients, such as
poly(dimethyl siloxane)-triethyl amine system [14]. Fig. 5 shows the
dependence of permeability coefficients of H_2, CO_2 and Xe on the
applied pressure. Until one atmospheric pressure, no effect was found
for both hydrogen and xenon or carbon dioxide of which the solubility
coefficients differ by more than one hundred times.

The effect of thermal histeresis of PMSP membrane on gas
permeability and diffusivity.

It has been reported that the permeability coefficients of glassy

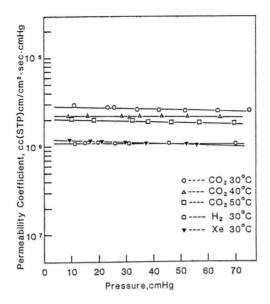

Fig.5. Pressure dependence of $\bar{P}$ for hydrogen, carbon
dioxide and xenon.

state membranes to gases are influenced by aging as well as thermal treatment. In the case of the PMSP membrane, decreasing the gas permeability with time or by thermal histeresis was especially remarkable. As one of the results, the effect of thermal histeresis of the membrane on the permeability, diffusivity and solubility co-efficients is shown in Fig. 6. It seems that the permeability coefficients at 30°C strongly depend on the highest thermal histeresis through which the membrane has ever experienced. The decrease of the permeability coefficients could be due to the decrease of the diffusion coefficients. This means that the change of morphology by the thermal treatment makes the membrane more dense.

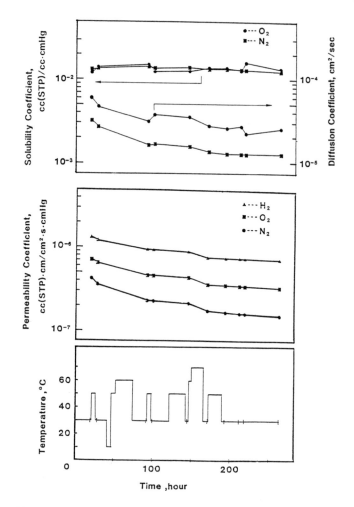

Fig.6. Effect of thermal histeresis of PMSP membrane on
Permeability coefficient of H_2, O_2 and N_2.

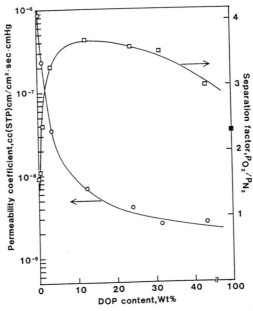

Fig.7. Dependence of DOP contents in PMSP membrane on the permeability coefficient of oxygen and the separation factor to nitrogen.

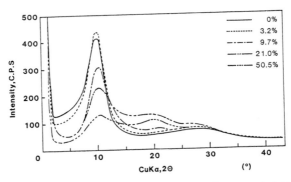

Fig.8. X-ray diffraction paterns of PMSP membrane containing DOP.

Modification of PMSP membrane by additives.

In an early stage of author's experiment, the PMSP membrane showed strong affinity to volatile materials. He considered that in addition to the thermal histeresis, the reason of unstable gas permeability is adsorption of volatile materials existing in atmospheres. Therefore, a modification of PMSP membranes by adsorption of additives in a higher temperature range such as 100 ° C. The small amount of dioctyl phthalate (DOP) , which is one of the most popular plasticizers for PVC, showed good results. Then dependence of DOP content on the permeability of oxygen and the separation factor to nitrogen was studied in more detail, as shown in Fig. 7. The permeability coefficient decreases with increasing DOP content. However, the separation factor reaches maximum at about 10 % of the DOP content. The separation factor of DOP membrane, which was supported in a filter paper, was found to be about 2.2. Therefore, the excess amount of adsorped DOP lowered the separation factor. The X-ray analysis of the PMSP membrane adsorped DOP is shown in Fig. 8. The adsorption of DOP decreased the height of sharp peaks at $10°$ of 2 θ and produced a new peaks at about $20°$, in more than 10 % of DOP content. It is considered that such contents of DOP effect the polymer structure, but the small amount of DOP is adsorbed simply on the intra-surfaces of the polymer segments. The high permeability coefficient of oxygen, 3-4 X 10^{-8} $cm^3(STP) \cdot cm/ cm^2 \cdot s \cdot cmHg$ at 30°C, and also the normal separation factor to nitrogen, 3.3, were easily obtained by such a simple modification of the PMSP membrane. The modified PMSP membranes showed stable in their permeability and seem to be prospective as membranes for oxygen enrichment.

References

1 H. K. Lonsdale, J. Memb. Sci., 1982,10, 77.

2. W. J. Ward III, W. R. Browall and R. M. Salemme, ibid, 1976,1, 99.

3. S. Asakawa, Y. Saito, M. Kawahito, Y. Ito, S. Tsuchiya and K. Sugata, National Technical Report, 1983, 29(Feb.)93.

4. S. M. Allen, M. Fujii, V. Stannett, H. B. Hopfenberg and J. L. Williams, J. Memb. Sci., 1977, 2, 153.

5. T. Masuda, E, Isobe, and T. Higashimura, J. Am. Chem. Soc.,1983, 105,7473.

6. T. Masuda and T. Higashimura, Accounts of Chem. Research, 1984, 17, 51.

7. T. Masuda, E. Isobe, and T. Higashimura, Macromolecules, 1985, 18, 841.

8. K. Takada, H.Matsuya, T. Masuda and T. Higashimura, J. Appl.Polym. Sci.,1985, 30,1605.

9. T. Nakagawa, T. Saito, Y.Saito and S. Asakawa, Maku (Membranes), under submitted.

10. V. Stannett, in diffusion in Polymers, J.Crank and G.S.Park, Eds., Academic Press, London, 1968, Chap.II.

11. T. Nakagawa, H.B.Hopfenberg and V. Stannett, J. Appl. Polym. Sci., 1971,15, 231.

12. H.J. Bixler, A.S. Michaels and M. Salame, J. Polym. Sci.,1963, A1, 895.

13. T. Nakagawa, Kogyo Kagaku Zasshi, 1969, 72, 723.

14. N. Minoura and T. Nakagawa, Kobunshi Ronbunshu, 1977, 34, 725.

Silicone Polymer Membranes for Air Separation

By C. L. Lee*, H. L. Chapman, M. E. Cifuentes, K. M. Lee, L. D. Merrill, K. C. Ulman, and K. Venkataraman

SILICONE RESEARCH DEPARTMENT, DOW CORNING CORPORATION, MIDLAND, MI 48686, USA

INTRODUCTION

The use of membrane technology for the separation of air or enrichment of oxygen has advanced substantially during the past ten years. Efficient membrane modules have been developed which enabled the system to be scaled up from laboratory scale to commercial size. The oxygen enriched air thus produced has been used in the medical field, such as heart-lung oxygenator, and industrial application. In the latter case, for example, injection of oxygen enriched air containing 25-35% oxygen to a furnace leads to higher flame temperature, reduces the volume of parasitic nitrogen to be heated and thus lowers the energy consumption.

Two types of commercial modules have been developed thus far for separation of air, that is, the hollow-fiber membrane modules and spiral-wound membrane modules. In both cases, the enrichment of oxygen is accomplished by passing a feed stream of air across a membrane that has a higher oxygen permeability than nitrogen permeability, leading to a higher oxygen content in the permeate stream. There are two key parameters which inherently control the performance of the membrane, namely, the permeability coefficient (P) and selectivity (α) defined as the ratio of the permeability coefficient of oxygen to nitrogen, e.g., α = P_{O_2}/P_{N_2}. Ideally, one would like to have a membrane that has both high P and α values. In reality, unfortunately, polymers

364

that have high permeability generally exhibit low selectivity and vice versa.

Silicone polymers are well known for high gas permeability. However, the gas permeability of silicone polymers depends markedly on the structure of polymer. Although there are a number of reports dealing with the permeability of various gases through polydimethylsiloxane in the literature,[1-6] the relationship between gas permeability and structure of silicone polymer has not been investigated systematically. To gain the fundamental understanding of such structure-permeability relationship, a project jointly funded by Dow Corning and Gas Research Institute was initiated.

Two classes of polymer were investigated, i.e. (a) unfilled silicone homopolymers and (b) silicone-organic block copolymers. In the former case, the effect of organic substituent on silicon and the structure of polymer backbone on the gas permeability were investigated. In the latter case, the effect of composition and block size of the block copolymer were studied.

In this paper we will present the relationship between permeability, the polymer structure and the economics of oxygen enrichment using these silicone polymers. Permeability of organosilyl substituted organic polymers also will be reviewed.

STRUCTURE-PERMEABILITY RELATIONSHIP

(A) Unfilled Silicone Homopolymers

The following three series of polymers were synthesized and their O_2 and N_2 permeability determined.

(1) Series A: $(MeRSiO)_x$, where R is C_nH_{2n+1}, $n = 1,2,3,6,8$ $C_6H_5, CH_2CH_2CF_3$ and $(Me_2SiCH_2)_x$

(2) Series B: $(Me_2Si(CH_2)_mSiMe_2O)_x$, where $m = 2$, 6 and 8

(3) Series C: $(Me_2Si-Z-SiMe_2O)_x$, where $Z =$ —⟨◯⟩— and —⟨◯⟩—

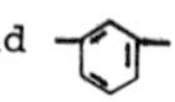

In the series A, polydimethylsiloxane, polymethylethyl-siloxane and polymethylphenylsiloxane were prepared by the base catalyzed ring-opening polymerization of the respective cyclo-tetrasiloxane in the presence of $ViMe_2SiOSiMe_2Vi$, using potassium silanolate as a catalyst. After the polymer had been fully equilibrated, carbon dioxide was added and the polymer was fractionally precipitated to remove the low molecular weight linear and cyclic species. For $(MeRSiO)_x$, where R is propyl, trifluoropropyl, hexyl and octyl, the polymer was made by ring-opening polymerization of the respective cyclotrisiloxane using lithium silanolate as a catalyst. Acetic acid was added to terminate the polymerization. Dimethylvinylchlorosilane was then added to convert the terminal hydroxyl group to a vinyldimethylsiloxyl group.

To prepare the silmethylenesiloxane polymer, 1,1,3,3--tetramethyl-1,3-disilacyclobutane[7] was first synthesized. The cyclic monomer was then polymerized with Pt/C as shown below:[8]

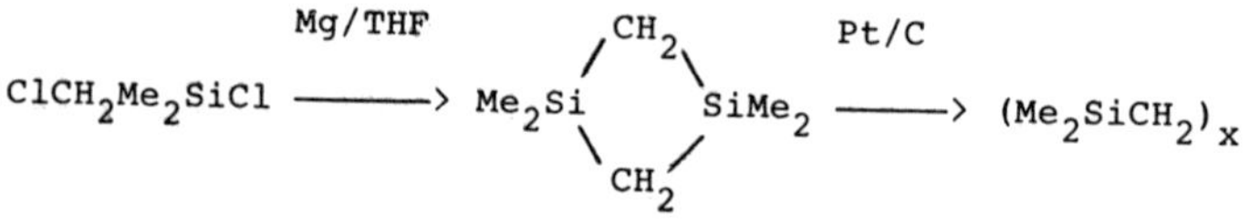

$$ClCH_2Me_2SiCl \xrightarrow{Mg/THF} Me_2Si\langle\overset{\displaystyle CH_2}{\underset{\displaystyle CH_2}{}}\rangle SiMe_2 \xrightarrow{Pt/C} (Me_2SiCH_2)_x$$

To prepare Series B polymers, the following intermediates were prepared first:

$$(CH_2{=}CHMe_2Si)_2O + HMe_2SiCl \longrightarrow (ClMe_2SiCH_2CH_2SiMe_2)_2O$$

$$CH_2{=}CH(CH_2)_xCH{=}CH_2 + HMe_2SiCl \longrightarrow ClMe_2Si(CH_2)_{x+4}SiMe_2Cl$$

These chlorosilanes were hydrolyzed to form oligomers,
$HO[Me_2Si(CH_2)_xSiMe_2O]_pH$, which were further condensed by azeo-
tropically removing water in the presence of a non-bond-
rearranging catalyst. The polymers were endblocked with
vinyl/group by reacting with $(Me_2ViSi)_2NH$.

Series C polymers were made by condensing
$HOMe_2Si$-⟨◯⟩-$SiMe_2OH$, which was made by hydrolysis of
HMe_2Si-⟨◯⟩-$SiMe_2H$.[3] The latter compound was made by the
"in situ" Grignard reaction shown below:[9]

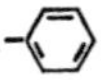

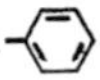

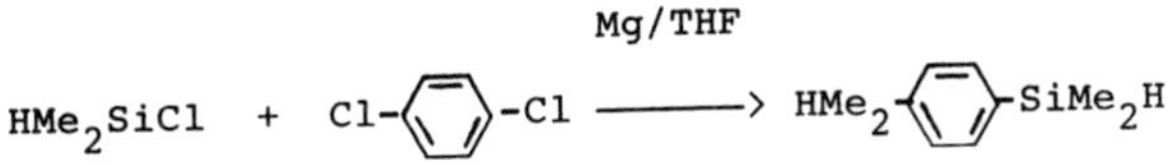

$$HMe_2SiCl \ + \ Cl-⟨◯⟩-Cl \ \xrightarrow{\text{Mg/THF}} \ HMe_2-⟨◯⟩-SiMe_2H$$

Both poly(tetramethyl-p-silphenylene-siloxane), PTMPS, and
poly(tetramethyl-m-silphenylene-siloxane), PTMMS, were prepared
in this manner.

Membranes were prepared by two methods, solvent casting and
compression molding. Solvent casting was used for membranes of
less than 0.003 inches in thickness. Poly(tetramethyl-p-
silphenylenesiloxane) membranes were made by this method.
Compression molding is suitable for membranes greater than 0.005
inches and has been utilized for most of this work. Crosslinking
of polymer was achieved by the peroxide vulcanization or by the
platinum catalyzed hydrosilylation.

Oxygen and nitrogen permeability were determined at 30°C
using a permeability cell of the variable-volume type.[10] Results
are summarized in Table 1. Polydimethylsiloxane (PDMS) exhibits
the highest permeability coefficient but the lowest selectivity
among the polymers studied. The effect of organic substituent,
R, on the permeability can be seen clearly in Series A polymers

TABLE 1.   GAS TRANSPORT PROPERTIES

POLYMER	PERMEABILITY[a] OXYGEN	NITROGEN	SELECTIVITY
Series A			
$(Me_2SiO)_x$	7.81	3.51	2.10
$(MeC_2H_5SiO)_x$	3.51	1.60	2.19
$(MeC_3H_7SiO)_x$	4.05	1.82	2.23
$(MeC_6H_{13}SiO)_x$	2.36	1.03	2.29
$(MeC_8H_{17}SiO)_x$	2.02	0.883	2.29
$(MeC_3H_4F_3SiO)_x$	1.83	0.833	2.20
$(MeC_6H_5SiO)_x$	0.256	0.0859	2.98
Series B			
$(Me_2SiC_2H_4SiMe_2O)_x$	4.23	1.86	2.28
$(Me_2SiC_6H_{12}SiMe_2O)_x$	1.82	0.810	2.24
$(Me_2SiC_8H_{16}SiMe_{2x}O)_x$	1.88	0.762	2.46
Series C			
$(Me_2Sim\text{-}C_6H_4SiMe_2O)_x$	0.756	0.319	2.37
$(Me_2Sip\text{-}C_6H_4SiMe_2O)_x$	0.141	0.0339	4.16
$(Me_2SiCH_2)_x$	0.910	0.3559	2.53

(a) $\bar{P} \times 10^8$ [cm³.cm/cm².sec.cmHg] wt 30°C

where the methyl group of the polydiorganosiloxane was held constant while the other substituent, R, was varied. When R is alkyl substituent, the permeability decreases by a factor of almost 3 as the size of R is increased from methyl to octyl. However, the selectivity, defined as the ratio of oxygen permeability coefficient to nitrogen permeability coefficient, remains more or less constant at about 2. Replacement of propyl group with γ-trifluoropropyl group causes further decrease in the permeability but the selectivity, again, remains about two. When R is aryl group, i.e. $(MeC_6H_5SiO)_x$, the permeability decreases 30 fold but the selectivity increases 40% with respect to $(Me_2SiO)_x$.

The effect of polymer backbone structure on the permeability can be seen in Series B and C in which every other oxygen atom in the polysiloxane backbone was replaced with alkylene and arylene linkage, respectively, while the methyl substituent was held constant.

Replacement of every other oxygen atom in $(Me_2SiO)_x$ with a m-phenylene linkage (PTMMS) caused a decrease in the permeability by at least one order of magnitude. When every other oxygen atom was replaced with an alkylene linkage (Series B) the permeability also decreased. Although the magnitude of decrease in permeability due to alkylene linkage increased with the size of alkylene linkage, the overall effect on permeability was not as pronounced as the phenylene linkage. The same effect was also observed in the permeation of drug molecule such as progesterone and testosterone through Series B and C polymers.[11] The introduction of phenylene linkage to the siloxane backbone apparently induced more rigidity to the siloxane chain than the alkylene linkage. When all oxygen atoms in $(Me_2SiO)_x$ were replaced with a methylene linkage, i.e. $(Me_2SiCH_2)_x$, the polymer chain became as rigid as PTMMS, resulting in a very low permeability.

The permeability of polytetramethyl-p-silphenylenesiloxane (PTMPS) was found to be the lowest and its selectivity (4.16) the highest among the silicone polymers investigated. The low permeability can be attributed to the crystallinity. PTMMS is an amorphous polymer, whereas PTMPS is a crystalline polymer (Tm = 148°C)

Figure 1 shows a plot of oxygen permeability coefficient vs. the selectivity for the O_2-N_2 system. Both organic polymers[12] and silicone polymers are included. The results indicate that polymers which have higher permeability coefficient generally exhibit low selectivity and vice versa, as represented by the curve shown in the figure. Ideally one would like to have a membrane which has both high permeability and selectivity, i.e., at upper right corner in Fig. 1.

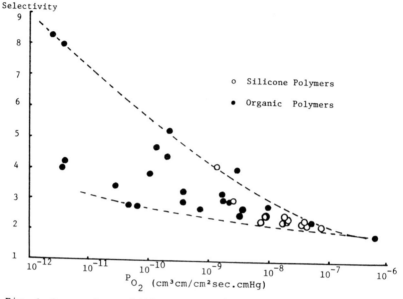

Fig. 1. Oxygen Permeability and O_2-N_2 Selectivity Relationship

The experimental data in Table 1 show that the presence of SiO bond tends to increase the permeability whereas the presence of p-phenylene group tends to increase the selectivity. Attempts were made, therefore, to synthesize a high selectivity polymer based on this principle, i.e. to incorporate more p-phenylene linkage into the polysiloxane backbone. Thus the following three polymers were synthesized and their permeabilities determined.

(a) $(Me_2Si\text{-}\langle\bigcirc\rangle\text{---}\langle\bigcirc\rangle\text{-}SiMe_2O)_x$

(b) $(Me_2Si\text{-}\langle\bigcirc\rangle\text{-}O\text{-}\langle\bigcirc\rangle\text{-}SiMe_2O)_x$

(c) Proprietary silicone resin

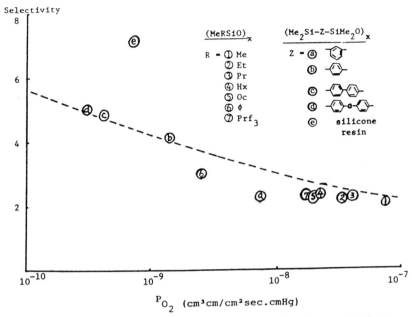

Fig. 2.   Effect of organic substituent and polymer backbone structure on the oxygen permeability and O_2-N_2 selectivity of silicone polymers.

Results are shown in Fig. 2. Polymer (a) and (b) gives higher selectivity but lower permeability than PTMPS, falling right on the curve shown in the figure. Encouraging result was obtained with polymer (e), which exhibits unusually high selectivity of about 7. Synthesis of other types of polymers based on the principle described above, therefore, might be fruitful for obtaining a more efficient polymer membrane.

(B)   Organosilyl Substituted Organic Polymers

Although the presence of siloxane bond in the polymers backbone is well-known to give higher gas permeability, it is less known that incorporating the organosilyl substituent to the organic polymer increases its gas permeability. Two examples

illustrating this point appear in the literature recently. These
are briefly reviewed below.

(1) Substituted Polystyrene

Kawakami and his coworkers[13] recently reported that the
permeability of polystyrene could be markedly increased by
introducing silyl or siloxyl substituent to the phenyl group.
Polystyrene has oxygen permeability coefficient and O_2-N_2
selectivity of 1.1×10^{-10} cm^{3}cm/cm^{2}sec.cm Hg and 3.8
respectively.[12] Introduction of p-trimethylsilyl group [i.e.
$(Me_3Si-\langle\bigcirc\rangle-CHCH_2)_x$] resulted in a significant increase in O_2
permeability but a decrease in selectivity as shown in Fig. 3.

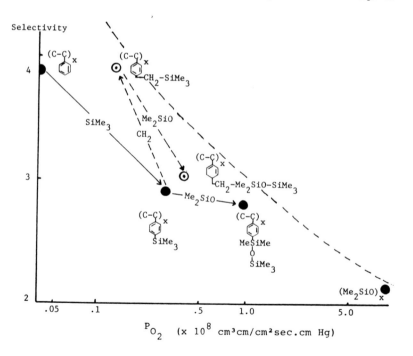

Fig. 3.   Effect of silyl and siloxyl substituent on the oxygen
          permeability of O_2-N_2 selectivity of substituted
          polystyrene.

Such an effect was enhanced by inserting a Me_2SiO unit, i.e.,

$(Me_3SiOMe_2Si$⟨◯⟩$-CHCH_2)_x$. Fifty fold increase in the oxygen

permeability over the unsubstituted polystyrene was obtained.

Its selectivity, however, dropped to 2.8.

Insertion of methylene unit, i.e.

$(Me_3SiCH_2$⟨◯⟩$-CHCH_2)_x$, on the other hand, caused a decrease in

the permeability but an increase in selectivity. Insertion of a

Me_2SiO unit to the latter polymer, i.e.

$(Me_3SiOMe_2SiCH_2$⟨◯⟩$-CHCH_2)_x$, of course, increased the permeabil-

ity in the same manner as $(Me_3SiOMe_2Si$⟨◯⟩$-CHCH_2)_x$. The effect

of silyl and siloxyl substituent on the permeability appears to

be dramatic. Thus, the modification of organic polymer using

silyl or siloxyl substituent is a powerful synthetic tool for

obtaining a high permeability organic polymer membrane.

(2) Substituted Polyacetylene

Trimethylsilyl substituted polyacetylene,

$(Me_3SiC=CMe)_x$, PTMP, is the most permeable synthetic polymer

reported in the literature to date.[14] The oxygen permeability

coefficient of PTMP is about 10 times higher than PDMS, but its

O_2-N_2 selectivity is lower than that of PDMS. Trimethylsilyl

group is apparently responsible for the unusually high oxygen

permeability since other organic substituents do not exhibit such

an effect[15] as shown below.

	$\begin{matrix}Me\\|\\(C=C)_x\\|\\\phi\end{matrix}$	$\begin{matrix}Me\\|\\(C=C)_x\\|\\C_5H_{11}\end{matrix}$	$\begin{matrix}Me\\|\\(C=C)_x\\|\\C_7H_{15}\end{matrix}$	$\begin{matrix}Me\\|\\C=C\\|\\SiMe_3\end{matrix}$	$\begin{matrix}Me\\|\\(SiO)_x\\|\\Me\end{matrix}$
P_{O_2} (x10^8)	.0746	.34	.384	61-83	7.8
α(O_2-N_2)	2.7	2.5	2.7	1.8	2.1

The magnitude of the triorganosilyl substituent effect on the gas permeability of substituted polyacetylene depends on the size of the alkyl radical of silyl substituent.[16] As the size of the alkyl radical is increased, the effect on the permeability diminishes drastically as shown in Fig. 4. The oxygen permeability of poly(dimethylhexyl)-1-propyne, for example, is three orders of magnitude lower than that of poly(trimethylsilyl)-1-propyne. It is interesting to note that the permeability-selectivity relationship of the triorganosilyl substituted polyacetylene follow exactly the same pattern observed in silicone polymers.

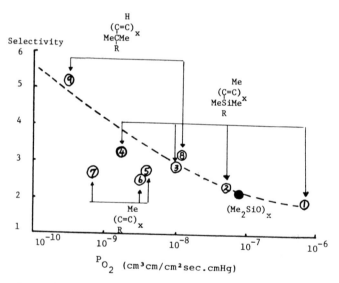

Fig. 4.   Effect of substituent on the oxygen permeability and O_2-N_2 selectivity of substituted polyacetylene.

$(RMe_2SiC=CMe)_x$: R = ① Me, ② Et, ③ Pr, ④ Hx
$(RC=CMe)_x$: R = ⑤ Hep., ⑥ pent., ⑦ phenyl
$(RMe_2CC=CH)_x$: R = ⑧ Me, ⑨ Et.

(C) Silicone-organic Block Copolymers

For a polymer membrane to be practical, it has to be thin
and strong in order to withstand the driving pressure of the
incoming gas stream. One of the weaknesses of the silicone
membrane is its poor mechanical properties such as tensile and
tear strength. One way to solve this problem is the incorporation of organic polymer blocks which reinforce the silicone
polymer. A number of silicone-organic block copolymers have been
reported in the literature. Recently, we have synthesized
various types of poly(dimethylsiloxane-co-ethyleneoxide-co-
methylmethacrylate). This will be reported in the forthcoming
paper. Their tensile strength can be increased to the range of
4,000-5,000 psi depending on the copolymer composition as
compared to about 50-100 psi for unfilled polydimethylsiloxane
elastomer . The effect of copolymer composition and block size
on the oxygen permeability is shown in Fig. 5. As the content of
polymethylmethacrylate (PMMA) is increased, the oxygen
permeability decreases but the selectivity increases. An
increase in the size of polydimethylsiloxane (PDMS) block causes
a significant increase in the oxygen permeability, whereas an
increase in the size of polyethylene oxide (PEO) block results in
a decrease in the permeability. The effect of block size on the
selectivity was found to be marginal.

ECONOMICS OF OXYGEN ENRICHMENT

One of the important commercial applications of membrane
technology is the oxygen enrichment of air. Increasing the
oxygen content of feed air to natural gas furnaces increases
flame temperature and efficiency. Two types of modules have been

(I)   Oxygen Permeability (P x 10^9)

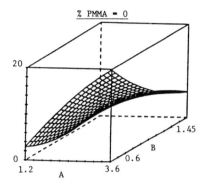

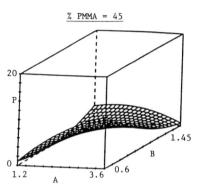

(II)   O$_2$-N$_2$ Selectivity (α)

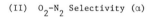

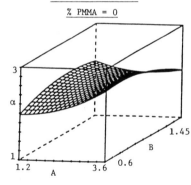

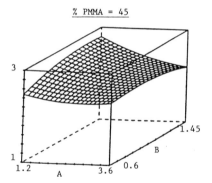

Fig. 5.   Effect of composition and block size on the oxygen
           permeability and O$_2$-N$_2$ selectivity of PDMS/PEO/PMMA
           copolymer.

           A,B = Molecular Weight of PDMS and PEO block
                 respectively (x10^{-3})

reported in the literature, i.e. the hollow-fiber membrane modules and spiral-wound membrane modules.

The economics of the air enrichment has been examined at Dow Corning using proprietary computer models. Input to the model are permeability coefficient, selectivity, membrane thickness, capacity, energy costs per kilowatt hour, and an estimate of the total cost per square foot of membrane area of the enrichment modules. From this data, the model yields optimum operating conditions necessary to minimize total costs. This is accomplished by calculating, at a given stage cut and pressure ratio, the volume of feed, volume of permeate and oxygen content of the permeate. From these values, the equipment required to handle such volumes can be determined, energy required to operate the unit, and maintenance/replacement costs are estimated.

The cost of oxygen enriched air can be expressed in terms of "equivalent oxygen," defined as the amount of pure oxygen required to mix with ambient air (21% oxygen) to produce the same quantity and oxygen content as produced by the enrichment process being considered. Enrichment plants are generally sized on the basis of equivalent tons of oxygen per day (ETPD). For example, if an enrichment process produces enriched "air" at 30% oxygen content, to produce 10 ETPD of oxygen, it has to produce 79.89 tons of enriched "air." In other words, 79.89 tons of permeate consisting of 30% oxygen, which is nearly 30 tons of oxygen, equates to only 10 equivalent tons of oxygen. By utilizing the equivalent ton basis, direct comparison can be made between a process such as cryogenics which produces nearly pure oxygen and the membrane enrichment process which may produce air containing as low as 30% oxygen.

In Figure 6, oxygen permeability coefficient is plotted vs
selectivity (P_{O_2}/P_{N_2}). Using the computer model, curves at a
given cost per equivalent ton are calculated. The silicone
polymers which have been studied to date are plotted along with
several organic polymers. The following assumptions are made:

a.   Membrane thickness in all cases is 1000Å (0.1μ)

b.   Membrane element cost is $6.24/ft². This is believed
to be a realistic estimate based both on Dow Corning's experience
and DOE reports authored Epperson and Burnett.[17] Note that there
is no difference in costs depending on the polymer considered.
This assumes that all polymers can be fabricated into membranes
with equal ease, an assumption which may be invalid. It further
assumes that the cost of polymer is negligible compared to cost
of fabricating and constructing the membrane elements. This
assumption is valid because at 0.1μ thickness, one pound of
polymer would produce approximately 50,000 ft² membrane.

c.   Electricity costs are $0.05/KWHR.

d.   Capacity is 10 ETPD.

e.   The lifetime of the system is twenty years, and the
lifetime of the membrane elements is five years. The capital is
depreciated over the lifetime of the system and elements using
straight line depreciation method.

The data shows that most of the silicone polymers examined
to date fall in the $30-40/equivalent ton oxygen range.
Interestingly, two of the polymers which were of interest due to
their high selectivities, $(Me_2Si-p-C_6H_4SiMe_2O)_x$ and $(C_6H_5MeSiO)_x$
give the highest costs. Thus, this data is valuable in that for
the first time a quantitative cost-permeability-selectivity
relationship is established.

OXYGEN PERMEABILITY

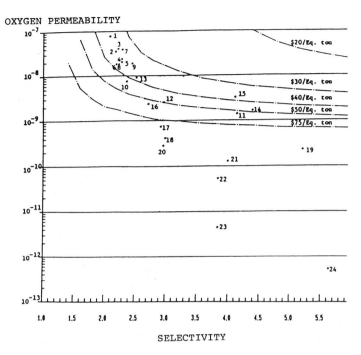

SELECTIVITY

Fig. 6. Cost-Permeability-Selectivity Relationship

1. $(Me_2SiO)_p$
2. $(MeEtSiO)_p$
3. $(MePrSiO)_p$
4. $(MeHxSiO)_p$
5. $(MeOcSiO)_p$
6. $(F_3PrMeSiO)_p$
7. $(Me_2Si(CH_2)_2SiMe_2O)_p$
8. $(Me_2Si(CH_2)_6SiMe_2O)_p$
9. $(Me_2Si(CH_2)_8SiMe_2O)_p$
10. $(Me_2Si \ m-C_6H_4SiMe_2O)_p$
11. $(Me_2Si \ p-C_6H_4SiMe_2O)_p$
12. $(PhMeSiO)_p$

13. $(Me_2SiCH_2)_p$
14. Polyphenylene Oxide
15. Polymethylpentene
16. Polyisoprene
17. Polystyrene
18. Teflon FEP
19. Polypropylene
20. Polyethylene (LD)
21. Butyl Rubber
22. Polyisoprene Hydrochloride
23. Polyvinylchloride
24. Polyvinylidene chloride

SUMMARY

(I)   Structure-Permeability Relationship

(1)   As the size of alkyl substituent in polydiorgano-
      siloxane is increased, the oxygen permeability
      decreases markedly, whereas the O_2-N_2 selectivity
      remains more or less constant.

(2)   Aryl substituent in polydiorganosiloxane or p-arylene
      linkage in the polymer backbone causes a marked
      decrease in permeability but an increase in the O_2-N_2
      selectivity.

(3)   Incorporation of organosilyl pendent group to organic
      polymers causes. pronounced increase in oxygen
      permeability but decrease in the O_2-N_2 selectivity.

(4)   Incorporation of organic block to the silicone polymer
      causes marked decrease in the permeability.
      Quantitative relationship between the permeability and
      the structure of PDMS-PEO-PMMA IPN elastomer was
      established.

(II) Economics of Oxygen Enrichment

      The cost of oxygen enrichment using silicone polymers was
analyzed using computer modelling. The cost for most of the
silicone polymers examined was found to be in the range of
$30-40/equivalent ton oxygen. For the first time a quantitative
cost-permeability-selectivity relationship was established.

ACKNOWLEDGEMENT

      The authors would like to express their appreciation to Dr,
F.W.G. Fearon of Dow Corning Corporation and Prof. S.A. Stern of
Syracuse University for technical suggestion during the course of
this work and to the Gas Research Institute of U.S.A. for the
financial support of this work.

REFERENCES

1. R.M. Barrer, J.A. Barrie and M.G. Rogers, Trans. Faraday
   Soc., 1962, 58, 2473.

2. R.M. Barrer, J.A. Barrie and N.K. Raman, Polymer, 1962, 3,
   596 and 605.

3. R.M. Barrer and H.T. Chio, J. Polym. Sci. C., 1965, 10, 111.

4. J.A. Barrie, J. Polym. Sci. A-1., 1966, 4, 3081.

5. W.L. Robb, Ann. N.Y. Acad. Sci., 1968, 146, 119.

6. M.S. Suwandi and S.A. Stern, J. Polym. Sci. Polym. Phys.,
   1973, 11, 663.

7. W. A. Kriner, J. Org. Chem., 1964, 29, 1601.

8. D. R. Weyenberg and L. E. Nelson, J. Org. Chem., 1965, 30,
   2618.

9. C. U. Pittman, J. Polym. Sci. A-1, 1976, 14, 1715.

10. S. A. Stern, T. F. Sinclair and P. J. Gareis, Mod. Plastics,
    1964, 42, 154.

11. C. L. Lee, K. L. Ulman and K. R. Larson, Drug Dev. and Ind.
    Pharm., 1986, 12, 369.

12. H. Yasuda, E. J. Brandrup and E. H. Immergut, "Polymer
    Handbook," Intersci. Pub., New York, 1966, Chapter V,
    p. V-13.

13. Y. Kawakami, H. Karasawa, T. Aoki, Y. Yamamura, H. Hisada
    and Y. Yamashita, Polym. J., 1985, 17, 1159.

14. T. Masuda, E. Isobe and T. Higashimura, J. Am. Chem. Soc.,
    1983, 105, 7473.

15. T. Higashimura, T. Masuda and M. Okada, Polymer Bulletin,
    1983, 10, 114.

16. K. Takada, H. Matsuya, T. Masuda and T. Higashimura,
    J.   Appl. Poly. Sci., 1985, 30, 1605.

17. B. J. Epperson and L. J. Burnett, U.S. Department of Energy
    Final Report, DDE/SC/40294, 1983.

The Structure of Gas Separation Technology

By D. L. MacLean

BOC GROUP, INC., TECHNICAL CENTER, 100 MOUNTAIN AVENUE, MURRAY HILL,
NJ 07974, USA

INTRODUCTION

There are four principal methods of gas separation:
absorption, cryogenics, adsorption (pressure swing adsorption),
and semi-permeable membranes. Emphasis will be placed on the
latter three technologies.

Absorption is a physical process (which can be chemically
enhanced) where a gas is selectively dissolved in a liquid and
subsequently recovered through the action of heat, pressure,
and/or another chemical. Absorption processes have found major
applications in the removal of acid gases such as CO_2 and H_2S.

Adsorption is a process in which one component of a gas
mixture is preferentially adsorbed on a solid surface which is
highly porous. These solid adsorbants may have surface areas of
1500 square meters per gram of material, mostly in micropores of
less than 40 Angstroms effective diameter. Adsorption processes
such as pressure swing adsorption where the gas is adsorbed at a
higher pressure and released at a lower pressure have been used
to recover H_2, CH_4, CO_2, CO, O_2, N_2 and other gases.

Cryogenic separations occur at low temperatures much below
0°C. Vapor and liquid phases are produced in order to take
advantage of the process of distillation. Cryogenic distillation
involves a series of vaporizations and condensations in which the
higher boiling species concentrate in the liquid phase which
flows down the column and the lower boiling components
concentrate in the vapor phase which moves up the column. Heat
is removed from the column at the top through a condenser while
heat is added at the bottom of the column through the reboiler.

Cryogenics is the predominant technology in the separation of atmospheric gases, methane from nitrogen, ethane and ethylene and is also used in hydrogen separations.

Gas separation by membranes involves the preferential transport of one constituent of a gas mixture across a relatively thin barrier. Membranes can be made of metal, glass, ceramics or polymers and usually are in sheet or tubular form. The membranes usually are hundreds of Angstrom up to microns thick and many times are supported on porous substrates of the same or different material. Membranes have been used industrially for most gas separations with some exceptions such as CH_4 from N_2 and for ethane from ethylene.

The expression "structure of the gas separation technology" may seem somewhat esoteric but all that is meant is the underlying components that make-up the technology. The similarities and differences among the processes will be discussed. Finally, illustrations of the synthesis of the technologies into new processes will be shown.

The structure is analyzed in terms of the process steps, the physics of the separation, the key mathematical expressions, the strengths and weaknesses of the processes and overall economic considerations.

TECHNOLOGY REVIEW

Membranes

Membranes which are thin barriers between feed and permeate gas streams have been used to selectively transport fluids since life itself. There have been however, two major technical advances that permit industrial use. The first was the research of Loeb and Sourirajan of UCLA where thin asymmetric membranes were developed from cellulose acetate. A thin, dense outside layer was formed on a thick, porous base layer. This allowed high flux as well as good selectivity. This same principle has been applied to many other polymeric systems. The second was the discovery by Tripodi and Henis of Monsanto that a composite

membrane of polysulfone could be formed by adding a thin coating
of a very highly permeable polymer which would act as a "tire
patch" to plug the holes in the thin, dense polysulfone layer
underneath. The thin layer was supported by a thick porous layer
underneath.

Membranes have been formed into separators by either winding
flat sheets into spirally wound modules or taking bundles of
hollow fibers and casting epoxy tube sheets on both ends and then
encasing the bundle in carbon steel shells with appropriate
entrance and exit nozzles.

Pressure Swing Adsorption

This technique uses a porous solid material such as a
zeolite, an aluminosilicate material, or a carbon molecular sieve
to preferentially adsorb one gaseous species versus others. The
adsorbent is packed in carbon steel vessels and a higher pressure
is used to adsorb while a lower pressure is used to desorb. Many
times several vessels are used with various steps besides the
adsorption step and low pressure regeneration step. These steps
might include pressure equalization between vessels, product gas
purging of the beds and backfilling the beds with product gas
prior to feed introduction.

An alternative process is temperature swing adsorption where
a higher temperature is used to regenerate the beds rather than
low pressure.

Cryogenics

Very cold temperatures, for example 4 K for helium or 77 K
for nitrogen are used to separate gas mixtures. The cold is
produced through compression followed by cooling by conventional
methods such as cooling water followed by Freon refrigeration and
then expanding through either a turbo-expander or a Joule-Thomson
valve. A portion of the cold gas is then used to cool the
incoming gas and this process is repeated until liquid is
formed. The liquid is separated by distillation or a series of
boiling and condensing stages where the vapor leaving the stage

is enriched in the lower boiling species and the liquid leaving
the stage is enriched in the higher boiling species. The product
can either leave the system as a gas or as a cryogenic liquid.

Absorption

The principal market for gas absorption is in the removal of
acid gases such as CO_2 and H_2S. The liquid solvent which
removes the acid gas can either use high pressure to enhance the
transfer of the acid gas from gas phase to liquid phase or
chemical means such as the reaction of the acid gas with the
solvent. For "physical" solvents, the solvent is regenerated by
reducing the pressure. For "chemical" solvents, regeneration is
accomplished by heating the solvent and driving off the acid
gas. The principal solvent type is the alkanolamines which were
discovered by R. R. Bottoms of the Girdler Corp. and patented in
1930.

THE ANALYSIS

Process Steps

The processes can be analyzed in terms of the energy source,
the gas contactor and the driving force (see Figure 1).

FIGURE 1

PROCESS STEPS

ENERGY	+	CONTACTING DEVICE	+	DRIVING FORCE	=	SEPARATION

It is important to note that compression is the source of energy
in cases of PSA, membranes, and cryogenics. In some physical
adsorption cases, high pressures are important; however, when
using common solvents such as MEA (mono-ethanolamine) and DEA
(di-ethanolamine) for acid gas removal (CO_2, H_2S), heat
provides the energy to reverse the reaction of the acid gas with
the solvent. Because compression is common to the three
processes, its proper utilization is the key to success in each
case.

The contactor in the membrane case is the membrane
obviously. The membrane is typically an organic polymer. There
has been some use of inorganic materials, for instance palladium
metal for hydrogen purification and ceramics for gas
separations. The inorganic materials usually require an
additional source of energy such as heat in order to obtain
reasonable permeation rates.

Adsorbents are the contactors in the PSA systems. Similar to
membranes these can be organic or inorganic materials. Carbon
based material such as carbon molecular sieve are used for
separation of air for nitrogen production while zeolites,
aluminosilicates are used for separations such as oxygen
production and carbon dioxide and water removal.

The contacting devices in cryogenic separation equipment are
the trays or packing in the distillation columns and also the
heat exchangers. These devices are metallic.

In most cases, the transfer process is mass transfer and the
driving forces are similar. The exception is heat transfer in
cryogenic heat exchangers which is dependent on the temperature
difference between the warm and cold fluids. The driving force
for membrane separations is the partial pressure difference for
each species from the high to low pressure sides. The driving
force for adsorption is usually considered to be the difference
in concentration between the gas phase and the stationary phase.
Distillation mass transfer is related to the composition
difference between the liquid phase and the vapor phase.

Fundamental Science

The fundamentals will be discussed under two divisions -
thermodynamics and rate theory. In cryogenic engineering, cycle
analysis, exergetic analysis and vapor-liquid equilibrium are
three areas related to thermodynamics. The cycle analysis
relates to topics such as Carnot, Rankine, and Ericcson cycles
which analyze the various steps of compression, expansion and
heat transfer with regard to thermodynamic efficiency. Little
cycle analysis has been done with membranes; however, cycle

analysis is required for PSA. The cycle analysis appears
different, however. In PSA, the key is to produce the maximum
amount of a specific purity of gas while not wasting pressure.
The cycle analysis involves optimizing the adsorption, balancing,
purging, backfilling, and regeneration steps and times.

Minimum work or exergy analysis utilizing the second law of
thermodynamics can be performed. As might be expected, one must
minimize the ΔT's in heat exchangers, compress as little as
possible and as reversibly as possible, and expand as much and as
efficiently as possible. Also, the separation should be as
incomplete as possible.

Surprisingly, phase equilibria are important in all the
separator technologies. Obviously, vapor-liquid equilibria are
important in cryogenic distillation. However, in PSA equilibrium
gas-solid isotherms are important in determining the feasibility
of separation, the time scale of the adsorption and the driving
forces for the mass transfer. Phase equilibrium is also
important for membranes in understanding the membrane formation
from a solution of one or more solvents or non-solvents.
Solution thermodynamics can also be used to predict the
solubility of various gases or vapors in the membrane.

In the rate theories, the expressions for mass transfer are
similar. See for example Figure 2. The driving forces were
discussed previously. One big difference between PSA and the
other processes is that PSA is an unsteady state process and the
concentration front travels as a wave and is not continuous as
are the temperature and concentration profiles of the others. In
other words, at any point along the contactor, there is no short
time variation in composition or temperature for cryogenics and
membranes. In contrast, for PSA, the composition does change
with time.

The rate transfer coefficients and areas for mass transfer
both relate to the physical chemistry as well as the economics.
In all cases, everything else being constant, it would be
desirable to minimize area which relates to cost. One way to do
this is to increase the rate coefficient. For membranes, the

FIGURE 2

Cryogenic

Heat Transfer $j_H = UA (T_{hot} - T_{cold})$

PSA

$$\frac{\partial C_{si}}{\partial t} = k (Y_i - f(C_{si}))$$

Membranes

$j_m = (P/\ell)_i A \Delta P_i$

where j_H =   heat flux
      U   =   overall heat transfer coefficient
      T_{hot} =   hot temperature
      T_{cold} =   cold temperature
      Y_i =   gas phase composition
      t   =   time
      C_{si} =   adsorbant phase composition
      k   =   rate coefficient
      J_m =   mass flux of species i
      $(P/\ell)_i$ =   permeability rate coefficient
      A   =   area
      P_i =   partial pressure

rate coefficient is called the permeability rate coefficient or just P over ℓ where ℓ is the thickness of the membrane and P is the permeability. Obviously, one wants to minimize ℓ while maintaining necessary strength. P can be thought of as the diffusion coefficient (D) of the gas in the polymer times the solubility of the gas in the membrane (S). Both D and S are functions of both the gas and the membrane material. Optimization occurs by proper selection of the gases, membrane and operating conditions of pressure and temperature.

The heat transfer coefficient is a function of the flow geometry, physical properties of the gas and the surface, and flow rates. For distillation, the higher the mass transfer coefficient and greater the driving force, the faster the transfer. For PSA, the theory is perhaps less developed; however, the size of the pores in the sieve are critical as is any interaction with the pore walls. The choice of the sieve for any gas separation is a key step.

The analysis of the process steps and fundamental science is followed by a critical assessment of the technologies. Special attention will be given to the strengths and weaknesses of the separation processes.

THE EVALUATION

Strengths and Weaknesses

Each of the four gas separation technologies will be analyzed with respect to their performance. Special attention will be given to product quality and general economic considerations.

Cryogenic Air Separation

The cryogenic separation of air has been evolving since 1895 when the basic patent was issued. In fact, one of the reasons cryogenic air separation is difficult to displace is its long history of improvements and cost reductions. In business language, cryogenics is far down the learning curve after the expenditure of billions of dollars. Aside from the high degree of process and equipment optimization, cryogenics is a clear choice where liquid products are desired either for their cold value or their transportability. Another key attribute is the economy of scale for cryogenic systems. Large systems scale up very well in terms of reduced unit costs in contrast to some of the new, non-cryogenic technologies that are modular. Low power costs are also achievable because of the high degree of heat integration and the proper choice of location and associated power grid. Gas or liquid purities are also produced in cryogenics which are higher than competitive technologies. Finally, certainly one of the most important features of cryogenics is the value of the co-products. One can obtain

oxygen, nitrogen, and argon in pure states as well as the rare
gases if desired.

One of the principal weaknesses of cryogenics is their
inability to economically scale down to very small sizes. This
can be circumvented by liquid delivery if the point of use is
economically accessible. The second disadvantage lies in the
highly integrated, enclosed systems which do not permit easy
handling of widely varying feed streams or allow for simple
retrofitting. For air separation, however, good control of the
feed system is possible since one is using atmospheric air and
controlling its flow rate through compression.

Membranes

Membranes' great strengths are their versatility and
simplicity. The same or similar membranes can handle a variety
of separations under varying feed conditions. Membranes can be
run at high pressures (e.g., 2000 psi) and at high differential
pressures such as 1650 psi. Membranes like cryogenics are
capable of high recoveries (> 90%). Membranes function best
where existing compressors are available for permeate
recompression or where the non-permeate is the product.
Membranes normally produce one product in the 90-99% purity range.

PSA

Pressure Swing Adsorption can recover gases such as nitrogen
or oxygen at relatively low energy costs which are comparable to
cryogenics when a single product is sought. Purities of 99 to
99.9% for nitrogen and 93-95% for oxygen can be obtained.
(Argon, being harmless, is included in the stated purity.) PSA
has a low capital cost. Also, like membranes, PSA can be
supplied to remote locations and locations where equipment size
is critical. PSA and membranes also provide the customer with
equipment he can run if he so chooses. PSA and membranes are
excellent for separations such as hydrogen from hydrocarbons.

Absorption

Absorption technology is excellent for CO_2 and H_2S removal, especially at low feed concentrations. At high levels of CO_2, membranes or cryogenics (especially if liquid CO_2 is desired) are preferred. Membranes are especially useful as a precursor to an amine system to handle significant feed variations.

The characteristic advantages and disadvantages are listed in Figure 3.

FIGURE 3

SEPARATION TECHNOLOGY COMPARISON

	Strengths	Weakness
Cryogenics	High Purity Low Energy High Recoveries	Higher Capital Complex Integration
PSA	Medium High Purity Low Capital Low Energy	Lower Recovery Single Relatively Pure Product
Membranes	Flexibility High Pressures Simplicity High Recoveries	Medium Purity Possible Recompression of Permeate
Absorption	Removal of Gas PPM Levels of Residual Acid Gases	High Partial Pressure needed for Physical Solvents Low Partial Pressure needed for Chemical Solvents Low Purity of Acid Gas

While these can be argued, as I see the situation, cryogenics can
give very high purities at high recoveries at the expense of
capital. PSA has the dual advantages of low capital and energy
while sacrificing recovery. Membranes provide flexibility and
simplicity while getting high recoveries of a medium purity gas.

Overall Economic Considerations

One of the common denominators of the three processes is the
capital versus energy trade-off. This is illustrated in Figure 4.

FIGURE 4

CAPITAL VS. ENERGY TRADE-OFFS

Cryogenics	Δ T vs Area
	Stages vs Recycle
	Cycle
PSA	Sieve
	Cycle Time
	P_H/P_L
	No. of Beds - Equalizations
	Purity vs Recovery
Membranes	Membrane Choice (P/1 vs α)
	T
	ΔP
	Purity vs Recovery

For cryogenics in the various heat exchangers, a constant
trade-off between temperature differences and exchanger areas
exists. Also, the height of the distillation columns or number
of stages directly affects the pressure drop and also can be
influenced by the recycle ratio. Furthermore, the choice of
cycle can be low power or low capital but seldom both. The cycle
is important in PSA also in terms of cycle time versus bed length
and number of beds and equalizations versus pressure or power

loss. The chosen sieve can usually either have fast uptakes and less selectivity (low capital - higher power) or the reverse slow and selective (higher capital - lower power). Also high purity at lower recovery can be exchanged for lower purity at higher recovery. Additionally, the choice of high and low pressure and their ratio influences power and capital oppositely. The last three choices also exist for membranes.

Also, temperature can be used with similar trade-offs in both membranes and PSA. More selectivity at lower temperatures but slower rate processes.

The question of products, single versus multiple, can be raised (Figure 5).

FIGURE 5

PRODUCTS - SINGLE VS. MULTIPLE

Cryogenics - Single - N_2 Generator
               Multiple - ASU (O_2, N_2, Ar, etc.)
               LNG, NAT'L GAS

PSA -   Single - Usual
               Multiple - N_2 + enriched air

Membranes -   Single - H_2
               Multiple - H_2/CO + H_2
                        - CO_2 + CH_4

With cryogenics, multiple products are the norm, e.g., an air separation unit (ASU), but there are single product systems such as nitrogen generators. For PSA and membranes, single products are common but sometimes multiple products exist, e.g., enriched air with nitrogen PSA or nitrogen generating membrane systems.

The product requirements also can influence the process. For example, higher gas purities and the need for liquid product strongly favor cryogenics. High pressures > 500 psi disfavor

PSA; however, low cost 99% purity gas is easily obtained from
PSA. For medium purity gas with fluctuating feed conditions and
high available pressure ratios, membranes are an excellent choice.

THE SYNTHESIS

Most of this article has been spent in analyzing and
comparing the four gas separation processes of PSA, membranes,
cryogenics, and to a lesser extent, absorption. It is, however,
possible and also desirable to synthesize processes which are
combinations of the four separation technologies; see Figure 6
for an illustration of some of the combined processes.

Membranes can be combined with PSA in either the process as
summarized by MacLean et al.[1] or the BOC (patent pending)
processes. In the first process, the low recovery of a
particular PSA can be upgraded by rejecting an impurity from the
vent gas after compression. The impurity tends to remain on the
non-permeate side of the membrane and the permeate product is
recompressed with the feed gas. In the BOC process the pressure
of the vent gas provides the energy for removal of impurities
across the membrane as the permeate while the purified vent gas
returns under some pressure to be used either as feed or
preferably equalization gas.

The BOC cryogenic process for argon recovery can also be used
after hydrogen selective membranes on ammonia purge streams.
This provides an excellent opportunity to remove hydrogen by
membranes and cryogenically produce liquid argon.

Perhaps the most intriguing process is the BOC HARP
technology (patents pending) for ammonia purge streams which
provides for membranes for hydrogen removal, PSA for methane and
some nitrogen removal and a single cryogenic column for liquid
argon (see Figure 7). This process combines the best
features of each gas separation technology into one process.

Finally, one can consider the addition of absorption to the
other technologies. Rectisol, the process of using methanol at
low temperatures to absorb carbon dioxide and then pressure

FIGURE 6

SYNTHESIS

	PSA	Membrane	Absorption	Cryogenic
Cryogenic	BOC Process	Cryogenic upgrade (1) & BOC Process	Rectisol	
PSA		PSA Purge Recovery (1) & BOC Process	H_2 PSA on absorber off gas	BOC Process
Membrane	PSA Purge Recovery (1) & BOC Process		NH_3 - PRISM "Gas" Membranes	Cryogenic Upgrade (1) & BOC Process
Absorption	H_2 PSA on absorber off gas	NH_3 - PRISM "Gas" Membranes		Rectisol
All	BOC Process			

Figure 7

HYBRID ARGON RECOVERY PROCESS

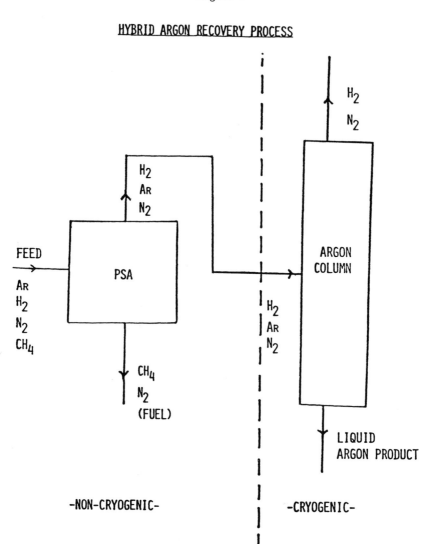

CONCEPT:

GAS PHASE SEPARATION OF CH_4 AND N_2

swinging the liquid to remove carbon dioxide, is a quite useful process. The PRISM^R separator process for hydrogen recovery from ammonia plant purge is a good example of the combination of aqueous absorption and distillation for ammonia removal and recovery and membranes for hydrogen recovery. Porous hollow fiber membranes can be used with solvents in the bore and aqueous stripping solutions on the shell-side in the manner suggested by Cussler and by Sirkar (2, 3) for gas absorption or liquid-liquid extraction. A hydrophobic hollow fiber is used and the volatile solute is trapped in the pores hollow fiber. It is this gas phase which provides the primary barrier or membrane. These membranes have been called gas membranes.

REFERENCES

1. MacLean, D.L. and Narayan, R.S., "PRISM ^{TM} Separator Applications-Present and Future", Hydrogen Energy Progress 2, 837 (1982).

2. Qi, Z. and Cussler, E.L., "Hollow Fiber Gas Membranes", AIChE J. 31, 1548 (1985).

3. Bhave, R.R. and Sirkar, K.K., "Gas Permeation and Separation by Aqueous Membranes...", J. of Membrane Science 27, 41 (1986).

Posters Presented at the Conference

Performance of Separex[R] Membranes in Hydrogen Service
W.J. Schell (Separex Corporation, Anaheim, USA)

Artificial Gills - A Method of Low Energy Underwater Oxygen
Recovery
P.C. Morgan and I.C. Risk (British Aerospace)

Gas Flow through a Graphitised Carbon Membrane
S. Butler (Diffusion Research Laboratory, Imperial College,
London)

Gas Separation Properties of Plasma Polymerized Perfluorocarbon
Membranes
M. Nakamura, S. Samejima and G. Kojima (Research Association
for Basic Polymer Technology, Yokohama, Japan)

Oxygen Enrichment through Fluorocarbon Thin Films
T. Kajiyama (Kyushu University, Hakozaki, Japan)

Use of Oxygen for High Temperature Combustion
A. Williams (University of Leeds)

Long Term Oxygenotherapy of Patients Suffering Heart/Lung
Inadequacies
Prof.Alexandrov (The Trade Delegation of USSR in UK, London)

Membrane Separation of Gases in Industries
Dr. Chekalov (The Trade Delegation of USSR in UK, London)

Oxygen Enrichment with Spiragas[tm]
T.E. Sulpizio (Fluid Systems, San Diego, USA)

Address Preceding the Toast to Joseph Priestley at the Priestley Dinner on 18th September 1986

By R. V. Giordano

CHAIRMAN AND GROUP CHIEF EXECUTIVE, THE BOC GROUP PLC

I should like to express my thanks to your Executive Committee for inviting me to pay this tribute to Joseph Priestley. People in the industrial gases business have particular reason to be grateful to him, and it is a happy coincidence that a BOC Priestley Conference should take place in the year in which The BOC Group is celebrating its centenary. A centenary is traditionally an occasion for celebrations and BOC has celebrated worldwide in a variety of ways.

A permanent and worthwhile reminder of our gratitude to Joseph Priestley is the establishment of the BOC Centenary Medal to be awarded to the keynote lecturer at BOC Priestley Conferences. Professor Patrick Meares is the worthy first recipient of this award. He gave a very fine lecture and made valuable contributions in the planning of this conference. You should also know that he is leading the campaign to make UK academia and industry more aware of membrane technology, a campaign now sponsored by the Science and Engineering Research Council.

This series of BOC Priestley Conferences came into being following the bicentenary celebrations in 1974 of Joseph Priestley's discovery of oxygen. It arose out of discussions between Dr Graham Winfield of BOC and our Chairman Dr Jack Barrett. Dr Barrett was then President of The Chemical Society, and has been Chairman of the Executive Committee for all the BOC Priestley Conferences so far. Dr Barrett's studies and considerations of many aspects of the life of Joseph Priestley have given him an enthusiasm for promoting an awareness of this great man that is itself a greater tribute to his memory than anything I can say this evening. Mr Chairman, you may be aware that we had a small number of busts of Joseph Priestley made. It seems to me to be highly appropriate that you should have one, and I have great pleasure in making this presentation.

401

From the beginning of the oxygen story, which we celebrate this evening, the separation of oxygen from air has grown to the extent that last year, in the league table of all chemicals produced in the USA, oxygen occupied fifth place, being surpassed only by lime, ammonia, nitrogen and sulphuric acid.

The first demand for oxygen to be satisfied appears to have been to make limelight early in the nineteenth century. In 1824 a Report of a Select Committee of the House of Commons led to a geographical survey of Ireland, undertaken by the Royal Engineers of the British Army. This survey required accurate observations of prominent features from a distance by viewing a light source on them by telescope. Experiments with flat mirrors to reflect the sun and curved mirrors reflecting light from oil lamps proved inadequate. One Lieutenant Thomas Drummond, trying to obtain a particularly bright light produced by incandescence, built a device using a piece of quicklime heated in a stream of oxygen by a spirit lamp. He had invented limelight. There are, of course, other claimants to this discovery, just as our French and Swedish friends would point to their compatriots involved in the isolation and identification of oxygen itself!

Limelight was 37 times as intense as the light from the best oil lamps of that time. On the evenings of 9th and 10th November 1825, Drummond's limelight was observed by his colleagues in Belfast, 66 miles away. Drummond went on to suggest improvements to his first model, with different configurations of the lime, and with other fuels such as hydrogen, but all using oxygen.

The intensity of limelight rapidly became celebrated and was proposed for applications such as lighthouses. An oxy-hydrogen limelight set up at Purfleet was capable, on a moonless night, of casting the shadow of a hand on a dark wall at Blackwall, more than 10 miles distant.

The theatre of the time was continually experimenting with lighting to produce new effects, so the brilliant limelight, warmer in tone and cheaper than electric arc lights, was attractive. Its first use is attributed to W C Macready at a Covent Garden pantomine in 1837. Macready charged 30 shillings per night for the hire of limelight - a considerable sum at that time. This covered the cost of an operator and of

hydrogen and oxygen, manufactured to order; neither was yet
available commercially. Oxygen was made by heating either
sodium nitrate or a mixture of potassium chlorate and manganese
dioxide in cast iron retorts, bubbling it through water, and
collecting it in gas bags. In the more prosperous theatres,
oxygen and hydrogen were stored in metal tanks at low
pressure. In all, it was a costly, cumbersome, and dangerous
way of going about things and this delayed the widespread use
of limelight for many years. But the market was there and the
time was ripe for a breakthrough in technology.

As is now well documented, the breakthrough came when the
Brin brothers persevered with Bousingault's barium oxide
process and set up the factory in Horseferry Road, Westminster,
to make cheaper oxygen and despatch it in cylinders in wooden
crates to all parts of the country. These were the humble
beginnings of The British Oxygen Company, 100 years ago.

Joseph Priestley was an honoured immigrant to my own
country in 1794. There was a very important outcome of the
celebrations at his home in Pennsylvania of the centenary of
his discovery of oxygen. I refer to the establishment of the
American Chemical Society. The admiration and respect that the
American Chemical Society has for Joseph Priestley is indicated
by the fact that the premier annual award of the Society is
called The Priestley Medal.

The 1984 Priestley Medallist, Linus C Pauling - recipient
of the 1954 Nobel Prize in Chemistry and of the 1962 Nobel
Peace Prize - has helped us to understand the significance of
Priestley's experimental method. I quote from Pauling's
Priestley Medal Address:

"In his scientific work Priestley exploited a new
technique, which opened up a new field of chemistry - the
technique of handling gases by collecting them over water
or mercury in a pneumatic trough. This innovation might
be compared to the 20th century introduction of X-ray
crystallography into structural chemistry or of
chromatography into analytical chemistry. By use of this
technique Priestley was able to discover 10 new gases and
to contribute significantly to the development of modern
chemistry."

These are the words of one of <u>the</u> great men of modern chemistry.
But Joseph Priestley was more than a scientist. He was a
theologian, teacher, and philosopher. In a paper issued by BOC
in 1974 to commemorate the bicentenary of Priestley's discovery
of oxygen, J G Crowther wrote "Priestley was a leading social
philosopher before he became a leading scientist. In 1768,
shortly after arriving in Leeds, he published his Essay on the
First Principles of Government. In it he wrote that the
government should interfere as little as possible with the
'lives, liberty or property of the members of the community'.
It was from Priestley that Thomas Jefferson took the most
famous sentence in composing the American Declaration of
Independence, published in 1776."

Today we are seeing Joseph Priestley from a distance in
time of 200 years. At a previous Conference you were reminded
by Dr C Beale of words used by his contemporaries on his
memorial. I wish to repeat them to you this evening:

"His discoveries as a philosopher will never cease to be
remembered and admired by the ablest improvers of
Science. His firmness as an advocate of Liberty and his
sincerity as an expounder of the Scriptures endeared him
to many of his enlightened and unprejudiced
contemporaries. His example as a Christian will be
instructive to the wise and interesting to the good of
every country and age."

I invite this company today to join me in the toast –
To the immortal memory of Joseph Priestley.